DESIGN
FOR A BETTER FUTURE

A guide to designing in complex systems

Routledge
Taylor & Francis Group

LONDON AND NEW YORK

John Body
Dr Nina Terrey

First published 2019
by Routledge
2 Park Square, Milton Park, Abingdon, Oxon OX14 4RN

and by Routledge
711 Third Avenue, New York, NY 10017

Routledge is an imprint of the Taylor & Francis Group, an informa business

British Library Cataloguing-in-Publication Data
A catalogue record for this book is available from the British Library

Library of Congress Cataloging-in-Publication Data
A catalog record has been requested for this book

ISBN: 9781138059801 (hbk)
ISBN: 9781315163918 (ebk)

Publisher's Note
This book has been prepared from camera-ready copy provided
by the authors.

CONTENTS – OVERVIEW

04

Design challenges

Dives deeper into the methodology and shows how it is embedded in practice. Explains how the Design System is tailored to various client audiences, approaching each design challenge in a unique way.

05

Core expertise

Describes the highly skilled teams that come together to meet each design challenge, and the ten core areas of expertise required to handle the complexity encountered in the design process.

06

Tools and techniques

Surveys the tools and techniques applied to each design challenge, offering evidence for how value is created through the design work.

Outlook

A look at how the methodology fits into the bigger picture, and the future for design in complex systems.

CONTENTS – IN DETAIL

LIST OF FIGURES

FOREWORD

In this century we enter a new unique phase of development. After agricultural and industrial revolutions, humanity is experiencing a different type of revolution. This time the drastic changes are not only augmenting our physical capacities, but also our cognitive system. New technologies have brought along cognitive innovations and opened new dimensions and capabilities that are creating new innovations, some of which are altering human values and even challenging some ethical norms.

Humanity is entering new frontiers with additional wicked challenges, and most of these new challenges cannot be understood without the context of systems, and in particular complex systems. These complex systems have dynamics and many intertwining feedback loops that are unprecedented. There are no more single answers or solutions. In open-ended human challenges, each challenge needs to be considered within its systems of interacting elements and feedback loops. In addition, the interacting components create unpredicted and emergent phenomena.

This book serves as an important blueprint for creating interventions to address these 21st-century complex human challenges. Using design, in its broad meaning, it delivers methodologies and processes that are centred around complexity to create real public, shared or collective value. In addition, this book teaches us about the abstract notion of complexity and how it is not only a mathematical term but is critical to humanity.

The subject is the connections and methodologies that address different types of complex human challenges, such as health, energy, transport, the environment, food security, strong organisations, law enforcement, regulation, poverty reduction, equality and financial inclusion, among others.

Armed with mathematical, design and business backgrounds, the authors advocate for different types of interventions and give us prototypes and examples for addressing them. What is impressive is that these prototypes, although maybe not fully translatable to a new challenge without some modifications, have been applied and proven viable in the real world. This book is a unique manual for tested tools that enable wicked human challenges to be addressed.

The authors, rightly, look to design as the overarching methodology to address human challenges. Design is an intellectual branch of knowledge and formally started almost 100 years ago. The word 'design' has different meanings in different contexts, but in general, design is a process and set of actions aimed at bringing a human system to a higher performing state. These processes and sequence of actions must adapt to the state of the system under consideration. The authors address this point very well and provide categories for different systems. They also propose very interesting analyses of the human challenges as well as provide skills and tools to be used for problem solving.

Today, the broad context of challenges changes with social factors, and includes economic, political, environmental and technological factors. These multi factors make designing in a complex system a unique process. Complex systems comprise many interrelated parts that don't lend themselves to a reductionist approach, and thus 'wicked problems' involve ambiguity, complex interdependencies and a diverse set of stakeholders. The authors use theoretical models that are drawn from chaos and complexity theory, systems thinking, ethnography and human-centred design to create their methodologies. They point out that in addition to the needed technical and economic solutions, mitigating a wicked problem requires changing the behaviour of millions of people, and integrating knowledge from several disciplines; a daunting task.

The authors consider the Cynefin framework, which has been useful to understand the level of uncertainty in a system, and approach complex challenges by designing the aggregation of sets of pathways through the system. In addition, they point out how design considers fractals of experiences from those of the individual, the cohorts of individuals and the whole system. The methodologies they introduce work smoothly between these levels of zoom, and from deep human empathy to a high-level strategic perspective.

The conceptual framework includes models and theories that underpin the practice of design thinking and must be multidisciplinary, rooted in fundamental concepts from industrial design, information design, mathematics, business, marketing, graphic design, architecture, experimental psychology, anthropology, sociology, and the study of policy and law. In particular, when issues are complex, the connections between design and engineering are critical. Design is forward-looking and explores what can be. Engineering translates design solutions into realities. Together they encompass both imagining the future and building it. But such outcomes cannot be achieved in isolation. Several other disciplines, including economics, policy and aspects of the humanities, must be considered and integrated. That, of course, will enrich and enable the solutions, and also add complexity.

During the past century, several authors have described the design process. In the 1980s, Peter Rowe at Harvard GSD wrote a book on design thinking and Eric von Hippel at MIT Sloan School discussed 'user-driven innovation' and the steps for successful design. However, bringing the design process to address complex challenges requires modifications of the well-understood and practised design process or what sometimes is referred to as the 'design thinking' approach.

The iterative systems design process might start by framing the challenge as a system, and within that take a human-centred perspective. It creates collaboration and conversation among designers and engineers, as well as with persons from different disciplines, and, in addition, engages the stakeholder. This yields the needed qualitative and quantitative outcomes. Keeping in mind that transformative solutions require explorations and divergent thinking, these solutions are driven into practice through synthesis and, later, via creation of virtual and physical prototypes. These prototypes must seek a balance of desirable, possible and viable. Of course, we expect that the process must be context specific and that the process considers a whole system approach and uncovers relations and interactions, addressing root causes rather than symptomatic and episodic information.

The execution of the design systems requires several constituencies to focus on: intent, design, experience and expertise, thus, the authors deploy the term 'fourth-order design'. This design is done in a tensor environment embodying three vectors of breadth, time and depth, and allows zooming among the elements of the overall system and the human experience. This approach is essential for studying systems in general, and especially complex systems.

The authors have derived system design steps that create discrete and sequential lenses. These lenses are iterative, and must be considered holistically. They start with the envisioning and framing step that is essential for any design, but of particular importance for complex problems. A second step of conceptual design is to understand what, how, when, and to whom change may occur, and a third step is to create conceptual frameworks, strategies, policies, services or product interactions, and organisational designs. These are followed by the prototyping/ making and measuring steps. This division makes the current design thinking process a subset of the fourth-order design that addresses complex systems.

The authors identified the role of the designer as not only an active creator and a process manager, but also a leader of individuals and organisations, and especially at time of change. Critical changes take place as the system is improved, as well as during the transformations that occur as a consequence of the strategic driving forces – these are the critical times that require exceptional leadership that designers should be able to provide.

The tools are a fascinating chapter written from real-life experiences. There, we see how the elements of the system interact and feedback loops can be explored and examined. In addition, valuable comments on research methods and techniques are presented and explored. Methods that help embody innovations are also presented and can be useful for addressing systems as well as new businesses.

At the end, we all agree that we cannot let our daily chores and work dictate the right path to work on time-sensitive issues such as pollution, loss of diversity, water and food scarcity, ocean warming, human population, and many urgent issues such as the ones articulated by the UN's Sustainable Development Goals. This book will be excellent to use as a manual for mobilising our thinking and acting in the directions that will address root causes of human challenges.

Fawwaz Habbal
Executive Dean for Education and Research
Harvard School of Engineering and Applied Sciences
Cambridge, Massachusetts

ACKNOWLEDGEMENTS

Designing in complex systems is a relatively new, multidisciplinary endeavour, and we are indebted to the vision, commitment and deep expertise of the many design practitioners who have contributed to it.

This book draws on the collective intellectual capital developed over the past decade or so by the people of ThinkPlace Global, particularly our leadership team:

John Body	Founding Partner
Nina Terrey	Partner ThinkPlace Australia, Chief Methodologist
Darren Menachemson	Partner ThinkPlace Australia, Chief Digital Officer
Jim Scully	Founding Partner & Managing Director, ThinkPlace New Zealand
Leslie Tergas	Partner & Design Director, ThinkPlace New Zealand
Bill Bannear	Managing Director, ThinkPlace Singapore
Dean Johnson	Managing Partner ThinkPlace Kenya
David Ireland	Principal ThinkPlace Australia, Global Innovation Lead
David Amos	Chief Operating Officer

Many other design specialists at ThinkPlace have contributed their design thinking and words, and critically reviewed drafts of the various sections. We particularly wish to acknowledge Ledia Andrawes, Cam Berry, Hayley Cosgrove, Natalie Coyles, Leah D'Mello, Eliot Duffy, Abram El-Sabagh, Erin Entrekin, Sarah Forrester, Dane Galpin, Carlyn James, Mondy Jera, Wai Ko, Phil Kowalick, Kathryn Lee, Ben McCarthy, Amy McLennan, Steph Mellor, Charlie Mere, Lydia Mitchell, Kerstin Oberprieler, Louisa Osborne, Roberto Persivale, Rafaella Recupero, Aimee Reeves, Cate Shaw, Mark Thompson, Simon Wong, Rose Wu, Wen Wen Ye and Annette Zou.

Production

Editor	David Evans-Smith, Evans-Smith & Dando
Photography	© Beth Jennings 2014–16 (www.bethjenningsphotography.com). Pages iii, 6, 14, 16, 26, 30, 32, 37, 56, 61, 101, 129, 133, 204, 234, 237
Design and typesetting	Ben Fulford, Keep Creative (www.keepcreative.com.au)

HOW TO BUILD A TIME MACHINE...

Why design anything?

Whether we are talking about a new model of car, a piece of furniture or a system for delivering services like health or education, it's a question that is deceptively simple.

The obvious answer? To make it better.

But what does that really mean? More sustainable? More accessible? More elegant? More efficient? And better for who?

The world is changing and the speed at which that change takes place has intensified. But can we really say that without intervention – without designers and design – it is getting better? That is why you – the person holding this book – are so important.

'Better' exists in the future state and – at our global design consultancy, ThinkPlace – so do we. Whatever else we are working on, we are always also designing a time machine. Our work often begins by mapping and understanding a current state that is problematic and then identifying a desired future state that those we are collaborating with wish to reach.

It is design for a better future.

This book mostly tells the story of how we go about getting there. How we do what we do. It is a methodology we've developed over more than a decade, combining an expert appreciation of systems complexity with the maturing fields of design thinking, innovation and human-centred design.

The result represents a roadmap for change. It is an instruction manual that allows the user to rethink and remake systems and services, cultures and organisations.

But before we lead you into the 'how?' let us first spend a moment reflecting on the 'why?' If not built upon the right intent, change will be pointless. Harmful even. After all, increasing the efficiency of an unfair or unethical system only serves to multiply pain and suffering.

That is why intent matters. At ThinkPlace we have set ours ambitiously, by adopting the United Nations' 17 Sustainable Development Goals as our guiding principles. Every project we take on aims to 'move the needle' on at least one of these goals. They include world-changing objectives like tackling hunger, ending poverty, addressing climate change and building strong economies and institutions.

We're going to be pretty busy.

What is your version of 'better'? It's a question we encourage you to ask as you progress through the different stages of this book, and to ask of the people you collaborate with. Without it, your full impact as a designer will not be realised.

Complexity

Having covered 'why?' we will tackle 'where?' Where does our design take place? In complex systems that have a set of defining characteristics that can make problems seem intractable and action appear daunting. For some this complexity can be demotivating. For us it is the opposite.

The defining characteristic of a complex system is, of course, that it is complex. A community of people is a complex system. So is an employment system, a health system and an economy. Within and between these systems are many different parts, connections and, perhaps most importantly, perspectives.

To achieve positive change within a complex system we need to bring these different perspectives together and guide them, as a group, towards creating something better. We need people who have accountability in the system and can set direction, people who experience the system (such as users), and people who know how the system works

– and we need people to guide and broker the process. What all these people have in common is a desire to imagine and shape a preferred future.

If you're one of these people, this book is for you. You may be faced with challenges affecting policy, strategy, programs, services, products or organisations. In complex systems these challenges are themselves complex, involving ambiguity, uncertainty, overlapping dependencies, a diverse set of stakeholders, and emerging, unplanned developments.

This book is intended as a resource to help you apply design thinking to these complex challenges. It's about fourth-order design – designing in complex systems and environments. (The four orders of design are explained in Chapter 2, 'Operative concepts'.)

At ThinkPlace, we've pioneered the application of design thinking to complex challenges. We regularly work with governments, organisations and communities in numerous countries,

creating effective change and public value in a systemic way. Our methodology – the Design System™ explained in this book – has evolved through both theory and extensive practice since we began in 2005. By sharing it we aim to help individuals, organisations, communities and nations achieve sustainable improvements in their complex systems.

What to expect

This book explains our design approach, methodology, and tools and techniques. While the subject matter is specific to the design industry, we've sought to make the language and concepts applicable and accessible to everyone who works with complex challenges. It provides a first look at the Design System methodology, which continues to be applied and adapted across a breadth of challenges such as health, energy, transport, the environment, food security, strong organisations, law enforcement, regulation, poverty reduction, equality and financial inclusion.

Our goal is not only to showcase design challenges and techniques but also to describe the art of design. The content of this book responds to the following questions: What makes this approach to design intellectually rigorous? What do we mean when we say we 'co-create value'? And more broadly, who relates to this value? Who can stake a claim to it? How is it that design thinking can maintain a powerful balance between reflecting the very fabric of people's lives and yet still challenge their assumptions?

This book is not just meant to be read – go out and apply it. We find it can be difficult to illustrate what we do without demonstrating it. We encourage experiential learning, embracing failure as a way forward. The way we see it, fearing failure stifles creativity. Innovation and failure interact with each other in productive ways. Therefore, don't think of this book as a guaranteed path to success. Think of it as a companion to failing early and often. In doing so you are growing your capacity as a designer.

The book you hold in your hands reflects our current thinking and practices, but we are constantly growing and evolving as designers and our methodology adapts accordingly. This book is a cross-section of our work, a living exhibition of the multitude of ways we apply our Design System to complex systems.

What you can expect from this book is an introduction to a way of thinking and acting that will allow you to work in complex systems to achieve scalable, sustainable change. It's about being both creative and analytical. Our approach to design is different, and we're prepared to defend it because we've seen it work time and time again. At ThinkPlace, we're intellectual and discerning about design, but we have a bias towards action. Our methodology gives us the creative confidence to say that where we lack understanding, we know how to go deep and learn more.

Use this book to get a glimpse of what we do, and then go and apply it. Design lives by practice.

About ThinkPlace

ThinkPlace is a global strategic design consultancy with deep expertise in design and innovation. We work on challenges that matter, where there is collective, shared or public value to be created. We seek to create value for societies, their economies and for the environment. We work simultaneously at the broad system level and the deeply human experience level. We bring diverse stakeholders together to have constructive dialogues. We dislike compromises, striving for simplicity – the other side of complexity – for all stakeholders.

We are fourth-order designers, which means we design in complex systems and environments. We experiment to find new ways of tackling intractable challenges.

Creating new and better futures is a team effort that includes those who drive change and those who will experience change. We engage with the complexity and the messiness of real human experiences — how people live, work and play, how they experience products, services and regulations, and how they navigate the systems they interact with, such as health, education, energy, economic, regulatory and social welfare systems, to name a few.

This gives us an understanding of the depth of human experience. We also map the breadth of diverse stakeholders that shape the system that people experience. We uncover and map their intentions for the system.

We use this understanding to generate innovation that can meet multiple needs because it is grounded in real human behaviour and broad system understanding. Whether we're working on a large-scale transformation, or designing a small-scale but meaningful change, we work from insight to impact, empowering people to create change that matters. We transform systems to work for people. Design is only as good as the experiences that are generated.

We thrive on chaos, ambiguity and seemingly unresolvable dilemmas. In uncertainty, we look for insight and opportunity. We derive deep satisfaction from working with clients and partners to understand the world they are living in, to re-perceive their reality, to define preferred futures and to design courses of action towards those desired futures. We design strategies, services and organisations to effect change. We build design and innovation capability in communities and clients. We have a strong core of expertise in collaborative human-centred design, innovation and complex system transformation. We combine these core strengths with relevant approaches such as gamification and big data analytics. We also experiment to find new ways of tackling these challenges and achieving transformation, to build new insight, and to ensure our capabilities evolve and stay relevant.

We have the courage and optimism, as well as the capabilities and reach, to design for positive impact at scale.

To deliver on this, we put people at the centre of our design process. We are pioneers in the field of co-design.

John Body
Nina Terrey

ThinkPlace Global
Canberra, Australia

UNCERTAINTY / PATTERNS / INSIGHTS

CLARITY / FOCUS

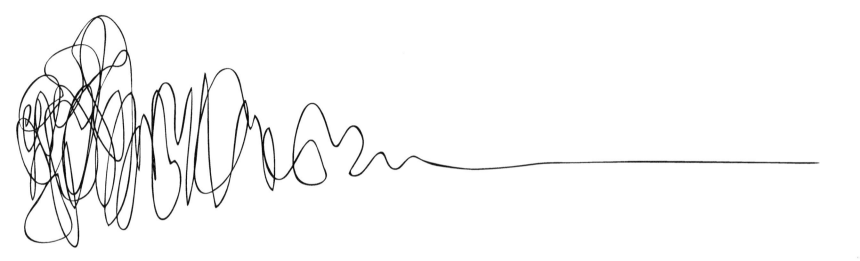

RESEARCH

CONCEPT

DESIGN

01

INTRODUCING DESIGN IN COMPLEX SYSTEMS

About complex systems, co-design and innovation, and why they matter

CHAPTER 1
INTRODUCING DESIGN
IN COMPLEX SYSTEMS

Design in the context of complex systems is both an art and a science.

It is an art because it's not possible to be prescriptive. There are judgement calls to be made. Every project is different, and what works in one context may not work in another. Sometimes the designer needs to be firm, at other times flexible. These judgement calls improve with practice and time.

It is a science because it draws on knowledge in various areas: how to run projects, how complex systems behave, research techniques, data science, ethnography, human behaviour and more. The strength of the outcome comes from a combined field of different disciplines.

Our design approach, described in this book, is founded on the principles of co-design and innovation. They blur the lines between expert and novice, and work together to transform complex problems into elegant solutions.

This design approach empowers people to generate solutions that work for them, and the empowerment of people in the process is what drives the pursuit of innovative solutions. It's an approach we've developed and tested through hundreds of real-world projects in diverse countries and contexts.

In this chapter we introduce complex systems, the case for a design approach to working with complex systems, and the key concepts of co-design and innovation.

"IF WE WANT EVERYTHING TO REMAIN AS IT IS, IT WILL BE NECESSARY FOR EVERYTHING TO CHANGE."

GIUSEPPE TOMASI DI LAMPEDUSA
WRITER
1958

ABOUT COMPLEX SYSTEMS

Complex systems are unpredictable. They comprise many independent parts – individuals, companies, pieces of legislation, technology and more. While the parts are independent there are relationships between them, with large numbers of non-linear, interacting elements. Complex systems also exhibit emergence – something that unfolds over time, that we often can't see coming and can't plan for. They are dynamic, even organic.

For example, consider an employment system. It includes employees and job seekers, employers of all shapes and sizes, governments (labour market policies, unemployment support, tax system), insurers, and organisations that specialise in the system, such as employment agencies and work health and safety experts. Each of these is an actor in the system, in the sense that they participate in it, and each has multiple connections to other actors and elements of the system. Each of the actors in the system experiences different pathways through the system – different not only from the other actors, but different also as circumstances change, such as the way an employee's pathway changes if they lose their job. Sometimes these pathways will intersect with other complex systems, such as the health or justice systems, and the actors will experience multiple complex systems simultaneously.

It's a cliché to say we live in a changing and connected world. What's relatively new, though, are the speed, extent and drivers of the changes and connections. Just think of the disruption to traditional services, media and politics as a result of widespread internet access, especially the advent of smartphones and similar devices. Complex systems, which are characterised by multiple interactions and perspectives, now have more of both, and both are more powerful and immediate.

When our consulting firm, ThinkPlace, began in 2005, the iPhone was still two years from hitting the market. The term 'big data' had not been coined and only a few organisations were glimpsing the power it offered. Computing in the cloud was a new concept. Artificial intelligence and the internet of things were yet to have any impact.

We had seen large-scale terrorism events such as the attacks on the World Trade Centre in the United States but terrorism on a smaller, distributed scale was just emerging. Climate change science, renewable energy and electric vehicles were not mainstream. China's gross domestic product was less than one-fifth that of the United States. The human genome had just been mapped but at a cost of $100 million – with the results, for most people, unimaginable and unattainable.

Without smart mobile devices, information was not accessible or transferable the way it is today. Attitudes changed more slowly. Individuals had less power over governments and corporations. National boundaries were more clear-cut because online trade and information flow were at a fraction of current levels.

This small sample of the changes we have seen in the past decade or so highlights the dynamic context the world is designing in – a context that is increasingly connected, fast moving and prone to rapid disruption. The opportunities and challenges that emerge from this context defy solution through traditional approaches. They can't be addressed solely through analysis and breaking them down into parts because they are interconnected, with feedback loops that amplify or attenuate in unpredictable ways. Techniques we might use to calculate the stresses in a steel beam can't be used to address the big global questions: how can we reduce poverty, lift health levels across communities and nations, ensure the security of clean water for all people, or design stronger organisations? These are a few of the 17 Sustainable Development Goals adopted by the United Nations, and they typify the classes of global challenges that require a different way of thinking.

A key feature of these types of challenges is that there are many stakeholders involved, often with competing requirements. For example, some may value a rainforest as a place to live, others value it as a place to derive income, while still others value it for its biodiversity and impact on climate. Another feature of this class of challenges is that the parts of the system are interdependent, so a decision or change in one part affects the rest of the system, and related systems. For example, in the health system the ability to map a patient's genome results in a new industry, new treatment options, increases in demand in the health system, big ethical questions and implications for policy makers, regulators and insurers.

These are complex challenges because there are many different views and motivations that no single solution will satisfy. Some can never be solved, but they can be improved. They require multiple inputs to understand them, such as deep empathy for the individuals involved, canvassing of a wide range of perspectives, statistical information and operational data.

Narrow, individual solutions to complex challenges are likely to add up to a world that no one wants to live in. For example, urban sprawl and traffic congestion are the result of individual decisions by car designers, road builders, developers, house purchasers, employers and all tiers of government across multiple agencies. Each decision may be independently sensible but the combination of decisions adds up to unintended consequences including lost productivity in traffic, environmental degradation, poor aesthetics and social isolation.

APPLYING A DESIGN APPROACH

How do we work with the challenges of complex systems? How do we even define the problems they pose? No single person can know all this complexity, and reducing it to its component parts won't work because the solution relies on understanding the interdependence of the parts.

To meet these complex, multi-faceted challenges we need a new approach – a design approach. In the early 1990s Professor Richard Buchanan, who at the time was Dean of the School of Design at Carnegie Mellon University, argued that the practice of design is especially well-equipped to work with the 'wicked problems' of complex systems. He proposed that the expanding scale of design practice could be understood as four orders of design (Buchanan 1992):

1. Signs and symbols (graphic design)

2. Objects (industrial design)

3. Interactions (service design)

4. Systems and environments (complex system design).

This book is about the fourth order of design. (The four orders of design are explained in greater detail in Chapter 2, 'Operative concepts'.)

The practice of design offers a way to bring together multiple disciplines and perspectives, foster collaboration and innovation, visualise and test solutions, and drive the realisation of solutions. At every level, design thinking provides the tools needed to define the problem and improve the system.

But working with complex challenges necessitates an approach to innovation and design that goes beyond the repertoire of rapid prototyping, user research, iteration and visualisation. It requires a deep understanding of how systems behave and how the people affected by systems behave, sophisticated collaboration that extends well beyond consultation, and a combination of qualitative and quantitative research. Fourth-order design approaches complex challenges by designing the aggregation of sets of pathways through the system – by considering fractals of the lived human experience.

ABOUT CO-DESIGN

Co-design is a design approach where the designer acts as a broker for different perspectives rather than operating as the lone expert. It's not merely consultation but a fundamental collaboration that actively engages multiple perspectives in the design process. The sponsors, subject experts, end users, staff and stakeholder groups are involved from the start and throughout the project.

At the heart of co-design is the belief that people within an existing system bring important expertise and insight. No one person can see the whole system, but they can contribute great insights into the parts of the system they interact with. Co-design has two important by-products: first, it significantly increases the acceptance of the change among those most affected; and second, it expands the capacity of the people who engage in it, allowing more scalable changes in the future.

Co-design is particularly suited to design in the context of complex systems, where there are many different and often competing perspectives that must be integrated into a successful design. There are inherent risks when designing in complex systems; co-design significantly reduces these risks.

Co-design emphasises diversity at the start of a project, it doesn't shy away from it. For designers within complex systems, this is the strongest way to deliver an outcome that speaks to everyone involved.

ABOUT INNOVATION

In contexts that need new thinking, normative approaches don't work. Complex systems are non-deterministic, so deterministic methods don't work. Our approach is non-normative and non-deterministic – that is, it produces innovative solutions.

Innovation can be defined as change that adds value. It's not merely change for change's sake. In whose eyes is that value created? The value is for multiple stakeholders, which may include individuals, or particular entities or companies, but also includes the collective – society, the business community, and government and non-government organisations. Innovation adds value for all of these people.

In complex systems, value extends into the concept of 'public value'. This is about creating, adding and sharing value. For example, say a new banking product is developed for small farmers. Value may be created for the farmer, who better manages his finances and generates revenue; for the bank, which benefits from fees, investment and potentially greater stability in the sector in the long term; and for the broader community, because the farmer is part of an economy that benefits society.

An innovation can also be defined as something that outperforms the normative approach. In a world where the broad environmental context keeps changing, if you're standing still then you're going backwards compared to the rest of the world.

How is the broad context changing? When evaluating the external factors impacting an organisation or business, we consider five key areas:

- social factors – demographics, lifestyle, values, age distribution of the population and behaviour

- economic factors – interest rates, taxes, trade, entrepreneurship and employment rates

- political factors – elections, policy changes and significant events

- environmental factors – impacts on the ecosystem (water, wind, soil, food and energy) and sustainability

- technological factors – speed of development and uptake, transport, energy, communication, and research and development.

It's within these five key areas that innovation, complexity and design intersect.

CONCLUSION

To design in a complex system we must challenge the assumption that complexity is a barrier. Each design challenge – be it policy, strategy, program, organisation, digital, product, service or international development – seems complex at first glance. The key is to juxtapose the complexity of the challenge with a single, focused intent. This destabilises the idea that complexity is a barrier and helps us work towards teasing out a meaningful solution.

Even so, navigating so many diverse perspectives can be difficult, and it requires a collaborative, systemic approach. This is why co-design is critical. Co-design reduces risk and supports innovation; overcomes boundaries and minimises misunderstanding; and enables us to access the full potential of collaborative thinking, a potential that lies beyond the sum of its parts.

The Design System and accompanying methodology described in this book are underpinned by co-design principles.

02

OPERATIVE CONCEPTS

Our perspective on the shifting landscape of complex systems design

CHAPTER 2
OPERATIVE CONCEPTS

To set the backdrop for the Design System, it's important to first understand the conceptual framework behind the methodology. This chapter is dedicated to the operative concepts, models and theories that underpin the practice of design thinking.

We will define complex systems in greater detail by delving into the evolution of design, explain how to achieve exceptional design by exploring different perspectives and innovating, and take a deep dive into the application of design in a complex system, culminating in the generation of public value.

This chapter showcases how we have maintained intellectual rigour about designing in complex systems over the past decade. The conceptual framework we engage with is truly multidisciplinary, rooted in fundamental concepts from industrial design, information design, mathematics, business, marketing, graphic design, architecture, experimental psychology, anthropology, sociology, and the study of policy and law. The theoretical models we use are drawn from chaos and complexity theory, systems thinking, ethnography and human-centred design.

"THAT WHICH HINDERS YOUR TASK IS YOUR TASK."

SANFORD MEISNER
PROFESSIONAL ACTING COACH
ATTR.

FRAMING PROBLEMS AND WORKING WITH UNCERTAINTY

The designer in a complex system faces a daunting challenge, operating in an environment where the problems are many and difficult to isolate, with overlapping permutations and interconnections. A reductionist approach won't work. Instead, to work with complexity we need a whole-of-system approach to framing the problem and understanding the uncertainty in the system.

Wicked problems

Complex systems present 'wicked problems'. Design theorist Horst Rittel coined this term to describe a class of problems that have no easy or clear solution (Rittel and Webber 1973). Wicked problems involve ambiguity, complex interdependencies and a diverse set of stakeholders. Solving a wicked problem may require changing the behaviour of millions of people, and integrating knowledge from several disciplines.

To get an idea of the complexity of wicked problems and the dilemmas involved, consider a project we conducted in a rural African setting, where nearly three-quarters of the population were uninsured and had to pay for health services. What should we do when:

- a government's social health scheme has a negative public opinion and yet a deep sense of internal pride?

- the provision of health-care services must continue in the face of institutionalised issues?

- there is a culture of entitlement in which people don't want to pay (much) for a service they feel ought to be free?

- the government steeply increases the price of their product, with less than a few weeks' notice to customers, while the perceived value is low?

- income is so unpredictable and irregular that people are unable to pay a static, monthly lump sum fee – yet the penalty for defaulting is five times the monthly contribution and so prevents them from re-entering the system?

- customer-centric, affordable private sector options are more desirable, but the government prefers to continue business as usual?

How do we go about designing in such an environment? Where do we start?

Rittel insisted that solving a complex problem was almost entirely dependent on how the problem itself was framed.

Following Rittel, we understand that navigating complexity is about framing problems in the systemic context in which they occur. It's about how you think about constraints and interact with a problem. It's about a collaborative design process in which the designer brokers a whole-of-system perspective. Framing is how we find a path to innovation, because innovation emerges not from strict planning but from a preparedness to deal with uncertainty.

Levels of uncertainty in systems

Systems display different levels of uncertainty and the approach of the designer depends on that level of uncertainty. Uncertainty is driven by the number of independent agents in the system, the amount of control over those agents and therefore how predictable the overall system behaviour is. For example, in the health system the ability to map a patient's genome results in a new industry, new treatment options, increases in demand in the health system, big ethical questions and implications for policy makers, regulators and insurers. These may in turn result in constraints on the way genome mapping and related treatments are conducted, and could even spill over into other systems, such as employment. What begins as a single, definable change has flow-on effects and feedback loops that impact throughout and potentially across systems.

The Cynefin framework (Snowden 2002; Kurtz and Snowden 2003; Snowden and Boone 2007) is useful to understand the level of uncertainty in a system and therefore how to respond. This framework (Figure 1) identifies four challenges or domains – obvious, complicated, complex and chaotic – each of which requires a different response.

The first two challenges – obvious and complicated – are distinguished by the ability to predict the result of your action. The other challenges – complex and chaotic – require more delay in response. (There's also a central 'disorder' domain, which is the state of not knowing which of the major domains you are in.) Let's look at each of the four domains in turn.

First, there are obvious challenges. These are obvious problems that offer up a simple solution. You observe them, categorise them and respond. It could be that you need to organise a simple meeting in another city. You are okay to respond, and you can predict the result with some accuracy. It's enough to diagnose the problem, and apply the usual treatment to get a result.

At the next level are complicated challenges. These require analysis. For example, you could be working on a complicated machine, such as an engine. It is not easy to comprehend but if you pull the engine apart – that is, reduce it to its component parts – and then reassemble it according to the original specification it will behave predictably. The focus here is on analysis according to a known specification, or schema, from which you can trace back to the problem.

Figure 1: Cynefin framework

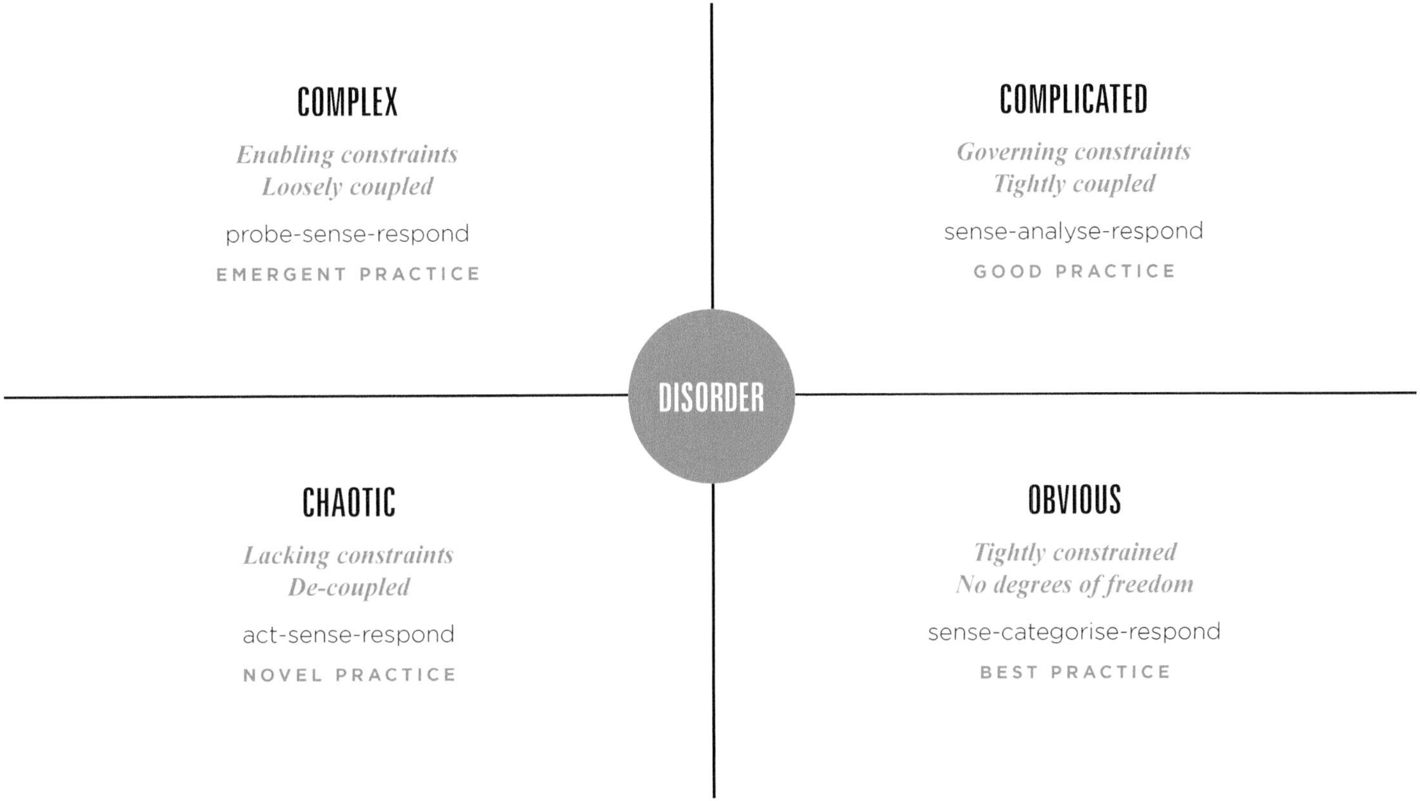

COMPLEX

Enabling constraints
Loosely coupled

probe-sense-respond

EMERGENT PRACTICE

COMPLICATED

Governing constraints
Tightly coupled

sense-analyse-respond

GOOD PRACTICE

DISORDER

CHAOTIC

Lacking constraints
De-coupled

act-sense-respond

NOVEL PRACTICE

OBVIOUS

Tightly constrained
No degrees of freedom

sense-categorise-respond

BEST PRACTICE

Source: Snowden 2002; Kurtz and Snowden 2003; Snowden and Boone 2007

Then there are complex challenges. These can't be readily solved and don't lend themselves to reduction because the solution relies on understanding the interdependence of the parts. The relationship between cause and effect is not immediate. The system is dynamic and the concept of emergence prevails because the components of the system interact and evolve, so it's difficult to predict or forecast outcomes. These are the challenges this book is about.

The fourth domain in the Cynefin framework is chaotic challenges. These are defined by their lack of predictability: you just don't know what's going to happen next, so there's no value in planning. Natural disasters and terrorism situations are examples of chaotic challenges. Interestingly, in a chaotic challenge the best response is to start with an action and see what happens, because planning will not increase the understanding of the challenge. Emergency services operate in this space but most people and organisations don't.

Each type of challenge warrants a different response. An effective response for one type of challenge will be ineffective if applied to a different challenge.

Design in complex systems often fails because people use methods meant for a more predictable world, without taking into account the higher level of uncertainty in such systems. We are comfortable with a predictable world so we convince ourselves that our challenges are predictable. We want solutions we've used before to work, without comprehending the unique particulars of a new situation. However, we are designing in an increasingly uncertain and rapidly changing world. Widespread, immediate and two-way information sharing through the internet, intensified by the advent of smart phones and similar devices, has enabled a range of disruptive shifts, including the sharing economy, empowerment of physically distributed groups such as social and political movements, and the increasing influence of social media compared to traditional mass media – not to mention disruptive technologies such as autonomous vehicles and genomics. At the beginning of the 21st century these terms and trends would have been meaningless to most of us.

The Cynefin framework also illustrates the value of co-design and innovation in situations involving complex problems, because the need to 'probe-sense-respond' aligns to the approach of co-design. In complex systems, a willingness to be experimental and explorative, to learn and evolve the solution, is critical to success. The problems or challenges someone brings you may appear 'simple'. The brief might even state exactly what needs to be done, and some people may respond by arguing simple problem-solving methods. When the designer in complex systems looks at a problem or challenge, however, they discern the underlying complexity of a so-called 'simple' problem and therefore argue the need to take a co-design approach.

Another way of looking at uncertainty in systems comes from philosopher Karl Popper (1991), who observed that there are 'clock problems' and 'cloud problems'. Clock problems can be divided into parts and clocks are utterly predictable, but cloud problems are indivisible, emergent and unpredictable; in other words, complex systems. A systemic problem that affects a culture is a cloud problem. This can refer to a personality, an era, a social environment, an economy, a built environment or a physical environment.

It's important to note that complex systems, though unpredictable, can possess emergent qualities that repeat and appear in ordered or patterned ways. The effect at an individual level or for a specific incident is unpredictable, but we can, with varying levels of accuracy, identify scenarios or trends that could appear across the system. For example, following Popper's analogy it's practically impossible to predict the existence

of or specific shape of a cloud, but we can identify different types of clouds that commonly emerge in different types of environmental conditions.

The Design System is specifically aimed at addressing complex challenges. However, before exploring the detailed question of how you apply design to a complex system, we will first explore what we mean by complex system design.

THE FOUR ORDERS OF DESIGN

The concepts of design thinking, design attitude and design awareness have been widely discussed since the 1970s, but it is a field that has not been easily reconciled or defined.

In 1992, Professor Richard Buchanan, who at the time was Dean of the School of Design at Carnegie Mellon University, said:

> Despite efforts to discover the foundations of design thinking in the fine arts, the natural sciences, or most recently, the social sciences, design eludes reduction and remains a surprisingly flexible activity. No single definition of design, or branches of professional practice such as industrial or graphic design, adequately covers the diversity of ideas and methods gathered under the label. (Buchanan 1992)

Buchanan proposed the concept of four orders of design (Figure 2), which encapsulated the growing view that the field of design could be applied at an ever-increasing scale.

The first order of design has a narrow focus, intending only to present something visually appealing or to communicate information. This field of design is literally millennia old, with evidence of early art on cave walls for aesthetic or communicative purposes.

Second-order design is the design of function. Evidence of this order of design can again be found in archaeological diggings with ancient tools and clay pots. First-order design and second-order design have been progressively refined and developed into the fields of visual design and industrial design, respectively.

Third-order design is the design of the interactions between people and objects. This field gained considerable impetus as technology developed and the need emerged to design human–computer interfaces.

Buchanan then proposed that design could be applied to an even bigger field: the design of societies, economies and environments. This is the fourth order of design, design in complex systems. Perhaps the defining characteristic of fourth-order design is that it approaches complex systems by designing the aggregation of sets of pathways through the system. In other words, this design approach considers fractals of experience.

The fourth-order designer identifies different cohorts of users who have similar experiences. Fourth-order design zooms between different fractals of the system – the individual, the cohorts of individuals and the whole system. A fourth-order designer can work smoothly between these levels of zoom, from deep human empathy to a high-level, strategic perspective.

Just as a designer of industrial products must understand material science – such as the physical properties of metals and plastics – the designer of complex systems must understand the properties and behaviours of complex systems.

As fourth-order designers we seek to understand how to create effective change and public value in a systemic way. We understand fractal behaviour – the self-similarity at different levels of zoom in the system. We understand bifurcation – that trends generally don't continue forever and instead they can rapidly change and branch out. We understand that a small change in initial conditions can make a big difference to the outcome. We know that complex systems comprise many interrelated parts that don't lend themselves to reductionist thinking. We know the power of working with parts of the system that are working well already, rather than hitting the complex system head on. And we understand the delicate balance between order and chaos: either extreme is unsustainable. The healthy space at the edge of chaos brings opportunity and innovation.

Figure 2: Four orders of design

ORDER OF DESIGN	FOCUS	FIELD OF DESIGN	DIMENSIONS
First order: **Signs and symbols**	Aesthetics and communication	Graphic design	Two dimensions
Second order: **Objects**	Form and function	Industrial design	Three dimensions
Third order: **Services**	Interaction	Service design	Four dimensions – includes time
Fourth order: **Systems and environments**	Whole system	Whole system design of societies, economies and environments through design of public policy, law, strategy, governance, organisations, systems and ecologies	n dimensions, multiple axes of concern, time, space, depth, experiences and relationships

EXPLORING ALL PERSPECTIVES

Implied in our design process is a way of looking at the world that embraces ambiguity and empowerment of people in a complex system. It promotes a style of working that recognises the value of curiosity and creativity, while also acknowledging that we rarely have all the data we need to drive perfect decision-making – especially when it comes to wicked problems.

The co-design approach canvasses different and competing perspectives to inform successful design. To ensure all perspectives are considered and thoroughly explored, we:

- conduct design research to understand the system

- hold strategic conversations to understand the intent

- incorporate the four voices of design to guide our thinking

- create a core design team to drive the design process

- base our methodology on the complex system tensor to develop a broader and deeper view of the various system perspectives.

In this way we ensure everyone has a voice in a true co-design process that creates lasting solutions for the greatest social impact and public value.

Co-design

Co-design, or participatory design, comes from a pragmatic and democratic view that users of a product, service or large public system have the experience and right to influence something that will affect their lives. Co-design seeks to engage all relevant actors in the design process, especially the non-experts in the system, to shape and influence the process and outcomes. This is how we know we are developing solutions that not only work, but that genuinely meet the needs of people.

Other problem-solving processes tend to transactionally engage users in arrangements that 'extract' their needs or requirements and only come back to 'test' for acceptance. These traditional practices don't embrace the power of user wisdom and experience.

The practice of collaborating with many people makes explicit existing knowledge of lived experiences and understandings, and creates new knowledge and understandings. By embracing the everyday person and inviting them, as non-designers, into the design process, the designer is all at once a designer, facilitator and educator, engendering environments that build trust, stimulate participation and educate participants in new modes of thinking, reframing techniques, and the tools to express and design new possibilities.

The users and designers embark on a creation process of ideas and solutions, striving to exchange an understanding of usage realities and solution possibilities, as users translate their needs into material changes to the interactions they have with services and products.

In co-design the designer ensures that ownership of the design challenge is shared among a collective of users and other actors, and that the eventual solution will be valued because it has resulted from their diverse experiences and ideas.

CO-DESIGN IS:	CO-DESIGN IS NOT:
A systematic approach to bringing user understanding into the design process.	Asking users what they want.
Searching for stated and unstated user needs to create value for all stakeholders, the individual and the collective.	Giving users what they want.
Prototyping possible solutions based on user insights and engagement with relevant experts.	Maximising value for one stakeholder at the expense of other perspectives.
Having an experimenter's mindset. Being prepared to fail early and often, and then to learn and improve.	A slow and laborious process to get it right the first time.
User understanding and engagement before there is a project and after implementation.	Consulting with users only after a design has been defined.
A process to rapidly generate and evaluate ideas, speeding up the development cycle.	Validation of pre-determined design solutions.
A capability that relies heavily on multidisciplinary teams to solve complex problems.	Disciplines doing their part of the design in isolation of others.
A win-win for all stakeholders – the individual in society, the whole society, the government, one company and whole industries.	Compromises and trade-offs.
A disciplined but flexible approach, which can adapt to ensure the intended outcomes are delivered.	A template-driven methodology that repeats in any context.
Iterative and non-linear; it can sometimes appear messy.	Rigid and ordered.

Design research

The fourth-order designer needs to understand (to the extent possible) the system within which they are working. Design research plays an integral role. Albert Einstein supposedly said, 'If I had an hour to solve a problem, I'd spend 55 minutes thinking about the problem and 5 minutes thinking about solutions.' Whether Einstein said this is doubtful, but the notion describes the essence of design research.

Many lines of enquiry exist to understand any given complex system. Design research uses as many lines of enquiry as possible to more completely understand the system at hand. Ultimately, all understanding will be flawed in some way, but the aim is to use practices that improve understanding and help to avoid three common mistakes.

The first mistake is to rely on only one method, such as a statistical view of the system, or an ethnographic view. Both have limitations, but combined they present a better picture.

Second, an inexperienced designer may fail to understand error in research. All techniques have error because they are trying to represent a complex system through abstractions. Error can come from looking at the wrong part of the system, asking the wrong people, following the wrong lines of enquiry or incorrectly applying the techniques.

Finally, an inexperienced designer tends to 'drink from a firehose' of data about the system rather than following a considered line of enquiry. The inexperienced researcher, for example, may run a series of unstructured interviews and have no framework for processing the answers to those questions.

Put simply, design research should ask three key questions:

- What do we need to find out?

- Who has the answers?

- How do we obtain the information accurately?

Research methods can include structured interviews, observations, ethnographic studies, surveys, transaction analysis, gathering national statistical information, researching international best practice and prototyping. The latter is a special form of research that involves learning by doing. Unlike traditional research approaches, design research works through iterative cycles of gathering information, considering it and building the design.

In chapters 5 and 6 we detail the expertise, tools and techniques that can be applied when conducting design research, since this is a core part of any design process in a project. In fact, the most powerful projects use a mix of all the above. There is a power in the combination of qualitative and quantitative techniques, and in the combination of big data (large-scale data sets) with small data (individual observations).

Strategic conversation

When seeking to understand the problem and devise possible solutions, strategic conversation is a powerful technique. We emphasise strategic conversation at several stages in the design process because it brings the best minds together to develop an understanding of the problem, identify the intent of the solution, generate alternatives and evaluate the preferred situations.

It may seem tautological to say that a good strategic conversation is strategic and a conversation, but these two terms deserve more explanation.

Strategic means to incorporate the breadth of different perspectives and design research. In healthy organisations the strategic conversation is ongoing, and senior leaders see their involvement as non-delegable. A strategic perspective includes stakeholder perspectives, the depth of human experience and how both might extrapolate over time.

Conversation is exploratory, non-combative and involves listening to and building on the ideas of others.

A good strategic conversation is designed to include a mix of perspectives that will bring different views of the system. We call these the 'four voices of design', examined in the next section.

Facilitation is a core competency for fourth-order designers because expertise is required to combine the different design voices in an authentic and non-contrived way to build a picture of the emerging future. The skilled facilitator guides the group over the course of the project, incorporating the design voices and design research to construct something far greater than any individual could achieve alone.

Four voices of design

The four voices of design (Figure 3) describe four perspectives that help us work within a complex system: the voices of intent, experience, expertise and design. These four voices carry the project through an end-to-end design journey, and ultimately help to move design from a place of empathy to one of execution.

The **voice of intent** sets direction, exercises authority and takes accountability for key decision-making during the co-design activity. This voice sponsors co-design, assuring alignment to the desired future customer experience.

The **voice of experience** is represented by citizens, industry, business, community members, users or consumers and their families, and service providers. This voice brings real, practical experience and needs to be engaged and collaborated with throughout the co-design process. These are the people who will experience the change first-hand.

The **voice of expertise** varies depending on the challenge. These are people with deep or specialist knowledge about the system. They may act as provocateurs or as sense-checkers. For an organisational challenge, this might include information technology (IT) teams, human resources managers, legal staff, and specialists in areas such as policy, marketing, communication and operations. This voice ensures that the change being designed is viable in operation and graceful in transition.

The **voice of design** acts as a broker for the other voices, ensuring all are heard in the right balance. This voice comes from practitioners that champion the creative, rapid, divergent and convergent design activity. It creates the plan to develop and achieve the strategic vision, balancing a desirable outcome within the constraints of what is possible and viable.

Figure 3: The four voices of design

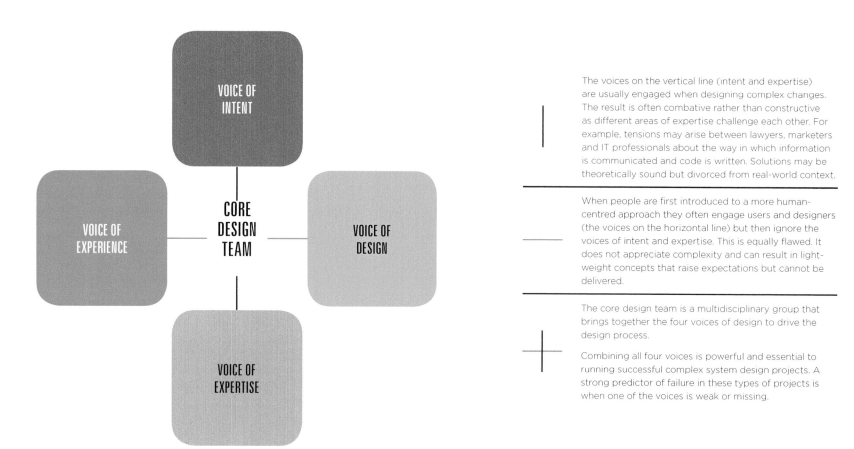

The voices on the vertical line (intent and expertise) are usually engaged when designing complex changes. The result is often combative rather than constructive as different areas of expertise challenge each other. For example, tensions may arise between lawyers, marketers and IT professionals about the way in which information is communicated and code is written. Solutions may be theoretically sound but divorced from real-world context.

When people are first introduced to a more human-centred approach they often engage users and designers (the voices on the horizontal line) but then ignore the voices of intent and expertise. This is equally flawed. It does not appreciate complexity and can result in light-weight concepts that raise expectations but cannot be delivered.

The core design team is a multidisciplinary group that brings together the four voices of design to drive the design process.

Combining all four voices is powerful and essential to running successful complex system design projects. A strong predictor of failure in these types of projects is when one of the voices is weak or missing.

Source: ThinkPlace and CoCreative Consulting

THE VOICE OF INTENT

...is about working from our spirit and our highest purpose. This voice invites us to dream big, bring our passion, and to aspire for better, more meaningful futures for all.

Principles

Craft a unifying purpose. Revisit the purpose in times of doubt or confusion. Deepen purpose over time. Differentiate, then integrate.

Methods

Purpose for the group and for each meeting. Share individuals' hopes and intentions. Define both focus and frame. Discuss frustrations and fears.

Traps

Dreaming over doing. Assuming unity without testing it. Groupthink. Failing to adapt purpose and strategies to changing context.

Gifts

Clear purpose. Personal growth. Deep engagement. Diversity within unity.

THE VOICE OF EXPERIENCE

...is about understanding with our hearts. This is the voice that reminds us to listen for pain and hope, to deepen our insight into the lives of others, to feel empathy.

Principles

Start understanding the system from the experiences of real people. Understand the context to create full solutions.

Methods

Context experts. Experience models. Experience prototypes. Journey maps. Personas.

Traps

Weighing lived experience over systemic understanding. Wanting to help without understanding impact. Settling for sympathy. Developing a 'we're helping the helpless' mindset.

Gifts

Focus. Empathy. Perspective. Grounded solutions.

THE VOICE OF EXPERTISE

...is about understanding with our heads. This is the voice of rational research and analysis, of selecting and tracking metrics, of measuring effectiveness.

Principles

Create a goal that's specific enough to measure progress. Treat everything as a hypothesis – and test it. See the whole system.

Methods

Map barriers and emerging opportunities. Invite experts to help the group learn more about the issues. Eliminate limiting factors.

Traps

Believing that rigorous analysis is more true than real-world experience. Getting stuck trying to understand the issue (analysis paralysis). Going around and around (death by debate).

Gifts

Rigour. Discipline. Methodical analysis. Systematic understanding.

THE VOICE OF DESIGN

...is about working with our hands. This is a creative, dynamic voice reminding us to work openly and collaboratively and iterate early and often to create solutions faster.

Principles

Test ideas early. Fail early and often. Quick rounds of brief feedback help more than one big round. Go with what's working.

Methods

Physical modelling. Storyboarding. Offering questions not suggestions in feedback rounds. Storyboards and scenarios to focus on HOW, not IF.

Traps

Losing steam when we hit hard realities and resistance. Incremental mindset. Experimentation over implementation. Losing track of the purpose and desired outcomes.

Gifts

Progress. Fast results. Momentum. Fast learning.

Core design team

'Everyone designs who devises courses of action aimed at changing existing situations into preferred ones.'

So said the scientist, economist and Nobel laureate Herb Simon (1988). His insight neatly captures the diversity and purpose of the core design team.

The core design team is a multidisciplinary, operational-level group, put together to drive the design process. It brings together the four voices of design – intent, expertise, experience and design – and the necessary diversity of skills and experience for the project. The team has a strong understanding of the intent and elevates it in every step of the design process.

The membership of the core design team should be maintained throughout the life of the project. It's not meant to be a body whose collective perspectives are comprehensive. Instead, a core design team should expect to enlist the assistance of, or at least have access to, people with various skill sets. The core design team provides regular updates to the steering group or sponsor overseeing the design objectives of the project. Ideally the team should comprise fewer than ten people so that it can maintain pace and energy.

The core design team should include people who are senior enough to make operational decisions but whose availability is not so constrained as to prevent the team from meeting regularly. Maintaining a diverse membership in the team makes it easier to move quickly and ensures that many different perspectives are available at each step. The people in the core design team should be carefully chosen. They should be optimistic – it's hard to design a positive future with pessimists who are not predisposed to imagining possible positive future scenarios. They should understand today's reality but also have an eye to the future and how things could be.

The core design team is not a representative group. It is small and nimble so that fragile emerging ideas can be developed into stronger hypotheses. But at some point a wider group needs to interact with these ideas. An extended design team, including a cross-section of the affected people, can be used for this wider engagement.

The complex system tensor

In addition to gathering the four voices of design, another way to engage multiple perspectives is to extrapolate the system spatially and temporally and to zoom deep into the system to understand the lived human experience. Put another way: look broadly, look ahead, and look closely. When we design, we have a unique way of doing this.

Strategy development often takes into consideration the breadth of stakeholders and trends over time. Human-centred design seeks to understand the deep lived experience. When designing in complex systems it's necessary to consider the three vectors: breadth, time and depth.

In mathematics, a vector is one dimensional, a matrix is two-dimensional and a tensor is three or more dimensions. We've found this thinking a useful way of describing, to those we work with, the strategic dimensions they should be examining when creating change in complex systems (Figure 4).

Figure 4: Complex system tensor

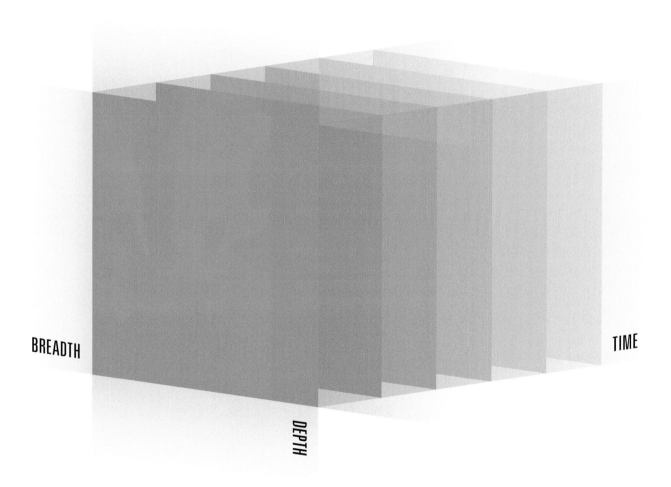

BREADTH

DEPTH

TIME

Vector of breadth (spatial)

When we design, the breadth dimension urges us to think about the broad context, such as social, economic, political, technological and environmental trends. These factors are beyond an organisation's control but strongly impact it. We also think about the organisation's context, including its market, clients, customers, competitors, partners, suppliers and the niche it occupies. Lastly, we think about the organisation itself and how it operates.

We can use statistics and surveys to quantify the system. This helps to understand the size of the system we are dealing with, including some of the flows and relationships in the system.

Vector of time (temporal)

We use this to extrapolate how the system may evolve. Of course, we can't predict the future, but we can imagine alternative possibilities by building scenarios that consider the critical uncertainties facing the system. This opens our thinking to how the future could be and therefore what approaches might make sense in those different worlds.

Every design project considers the evolution of big trends in today's world. For example, mobile devices now penetrate significant proportions of the community in developed and developing economies alike. Some organisations predicted the implications of this connection, embraced them and leapfrogged other organisations that lacked this foresight. The next big waves in technology and communications include the internet of things, artificial intelligence and the widespread mapping of the human genome. How will these changes impact an organisation? How can we change our current behaviour to capitalise on these changes?

Vector of depth (experiential and empathetic)

Strategists think about space and time, but designers in complex systems add one more important vector: that of depth. The depth vector allows us to zoom between the whole system and the deep lived human experience, expressed as a pathway through the system. We understand depth using ethnographic techniques, observing people, talking to people and asking them to record their experience. The depth dimension is essential to design but often overlooked. If we rely on broad perspectives and statistics alone, we don't get insights about how to improve the system. We lose empathy for the people in the system.

For example, an organisation with a client base of one million people may be encouraged by a report that 80% of its clients are happy with its service. However, let's consider the 200,000 people who reported they were not happy – in real terms, that's a lot of unhappy people. That big number contains stories of hardship, frustration and angst that, when understood, can significantly change an organisation's approach.

In every assignment, the complex system designer should ask whether they have considered all three dimensions, and if not, why not.

INNOVATING

As designers in complex systems we shouldn't expect to know everything at the outset or get fixated on a design solution. Instead, we need to embrace others' expertise and be ready to construct knowledge and ideas with the ecosystem of users, experts and leaders. With co-design as a foundation, our design approach uncovers potential solutions by emphasising:

- innovation as learning

- divergent and convergent thinking

- rapid iteration.

Innovation as learning

Our design approach moves through abstract and concrete stages to travel away from traditional mindsets toward new possibilities. The model shown in Figure 5 is a useful framework to understand learning as an innovation process, because it 'moves its participants between the concrete and the abstract worlds, and it alternately uses analysis and synthesis to generate new products, services, business models, and other designs' (Beckman and Barry 2007, p. 29).

Innovation can't happen without the evidence base that research supplies (observations). Making sense of the data collected during research is a process of framing and reframing to extract patterns and insights (frameworks). Once the user needs have been framed, the process moves to synthesising a set of design principles (imperatives) that must be met by the new concept that is being developed. Last in the sequence, the innovation process returns to the concrete by selecting viable concepts and testing them with users (solutions).

The essence of making something new is to have a constant desire to learn – to uncover, discover, create – that addresses the deep underlying issues and needs of people. This introduces critical activities in a project that value time toward observing, reflecting, sense-making, and acting with the intent to learn and gather more insight to make a better solution.

The participatory aspects of innovation as learning opens an exciting and unknown territory in project work, because the innovation process of learning will generate richer ideas and solutions than can be generated by an expert few. Embracing the mindset to learn opens the design process to endless possibilities and potential to generate real value. In complex public value challenges there is no single keeper of insight or invention – it is a co-discovery of learning and doing.

Figure 5: The innovation process

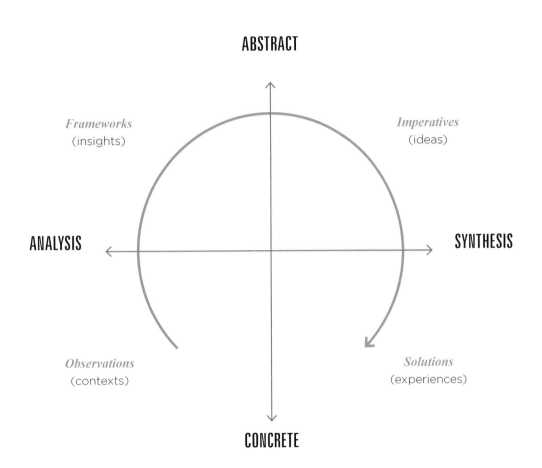

ABSTRACT

Frameworks
(insights)

Imperatives
(ideas)

ANALYSIS

SYNTHESIS

Observations
(contexts)

Solutions
(experiences)

CONCRETE

Divergent and convergent thinking

Psychologist JP Guilford (1967) drew a distinction between two forms of thinking: divergent and convergent. According to his definition, divergent thinking creates alternative theories and ideas while convergent thinking emphasises correct answers. Divergent thinking is critical in the creative process, whereas convergent thinking is particularly useful in the elaboration phase when it's important to discriminate and choose between alternatives.

Dr Tina Seelig, Professor of the Practice in the Department of Management Science and Engineering at Stanford University, explained divergent and convergent thinking by posing these questions:

> ?+?=10. What is the answer? There are infinite answers! That is divergent thinking – the way we ask the question opens up our thinking.
>
> 5+5=? It has to be 10. That is convergent thinking. There is a right answer. (Seelig 2012)

In the design process, divergent thinking encourages deeper, more original exploration than other approaches because it seeks to break free from constraints, existing perspectives and models. At the heart of divergence lies the art of discovering new problems and opportunities, of shifting paradigms to stimulate the invention process.

Convergent thinking uses focus and the prioritisation of opportunities to meet the needs and desires of users and other key actors. It creates compelling concepts with a high probability of success by working within constraints. American designer Charles Eames stated:

> Here is one of the few effective keys to the design problem—the ability of the designer to recognize as many of the constraints as possible and his willingness and enthusiasm for working within these constraints. Constraints of price, of size, of strength, of balance, of surface, of time and so forth. (Eames 1972)

Figure 6: Divergent and convergent thinking

Convergent thinking embraces constraints and drives prototyping, testing and validating solutions and modelling experimentation. Its forward-looking point of view uses success criteria to discriminate between multiple possible solutions to most effectively meet user needs and desires.

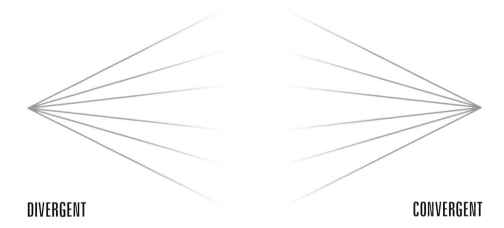

DIVERGENT   CONVERGENT

Figure 7: Tailoring the design process to meet the design challenge

Rapid iteration

The appetite to materially produce something from each stage of a design project is inherent in the designer's approach. As we move from divergent to convergent thinking, we embrace a rapid and visible materiality of the design process in terms of what we have learnt, what design possibilities exist and versions of the solution. We settle our thinking through visual representations, documentation, images, sound and creation of new knowledge. We capture quotes, insights, ideas, solutions, constraints and models. There is energy directed at making new versions quickly and iteratively. There is a high degree of comfort sharing very early ideas and concepts because there is shared ownership of the insights, ideas and concepts.

The paradox that design brings to any project is a slowing down, to go faster. By this we mean the exploration and innovation time is resisting the desire to converge on a solution, and by iterating and making representations of these stages, leads to highly insightful and meaningful solutions that make a difference to people and societies.

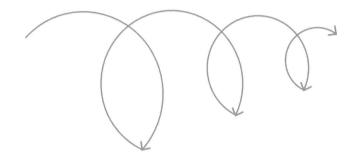

RAPID - ITERATIVE - INNOVATIVE

APPLYING DESIGN TO A COMPLEX SYSTEM

To realise the vision for a complex system requires an achievable design that is actually achieved. The designer needs a coherent way of unpacking the various elements and connections in the system to develop a coherent design, and an understanding of the levers available to effect change. Importantly, this understanding informs the designer's key role in the implementation of the design – that is, in implementing change.

There are four broad questions that can guide a line of enquiry when working with complex systems, particularly complex human systems:

1. What is the system there to do? That is, what is its purpose?

2. What is the experience of people interacting with the system?

3. What products, services and channels of interaction make that experience?

4. How are the products and services delivered?

Having worked through these questions in the current state, they can be turned into future state questions:

1. What should the system do? That is, what is its desired purpose?

2. What is the desired experience of people interacting with the system?

3. What is the best product, service and channel mix to deliver that experience? How can the products and services be improved?

4. How could delivery of products and services be improved?

These four questions provide a set of lenses for breaking down complexity while also thinking about an organisation holistically. They are the layers of design.

The four layers of design

The four layers of design (Figure 8) are:

1. strategy

2. experience

3. service

4. delivery.

This structure provides a coherent design because it has a strong user experience focus that is aligned to the strategic intent and sustainability of the design.

Figure 8: Four layers of design

The **strategy** or **intent layer** is about context, drivers, direction, outcomes, benefits and shifts.

The **experience layer** is about understanding the user journey, motivations and goals, and designing interactions. It thinks about how the experience will look and feel for the user.

The **service layer** describes the specific products, services and channels that the user interacts with. These might include payment services, information products, service charters and channel strategies.

The **delivery layer** is about technology, systems, data, staffing, roles, resourcing and processes.

The strategy layer

The first layer questions policy or strategy: what is the system there to do? Why do the transport, education, law enforcement, tax and welfare systems exist? Alternatively, if you are designing within part of a system, you will instead ask this question of separate functions. Why does the audit function, the compliance function or the monitoring function exist? The first question could be rephrased as:

- Why do we exist? What is our purpose? What value do we create?

- How do we create value?

- What do we do? What is our core business?

- Who are we? What are our identity, culture and values as a business?

- If we didn't do what we do, what would be lost?

The experience layer

The second question examines the human experience of people interacting with the system. This focus on experience is a defining feature of the design approach because it provides an integrative way to consider the system. Reductionist approaches tend to exclude human experience, which is why they are unsuited to complex systems. All the elements of the system combine to deliver an experience that is the total effect of (for example) a piece of legislation, an information system, a website, a call centre and a face-to-face service. A bad experience is one where there are unmet needs, which may or may not be apparent to the user. These unmet needs can come from irritants of all kinds, things that don't match people's preferences, or a user with a new set of circumstances.

Complex public system experiences are not like other experiences. A user may have several touch points with public systems throughout his or her lifetime, so it's important to understand the way a particular experience fits within a user's broader life journey.

For example, if a user gets sick, she will have a certain set of touch points with the health system. When the user goes to school, she will have a certain set of touch points with the education system. This unpredictable, adaptive route through complex systems is part of the user's life journey. When the user moves through the channels and processes of a single system, that's what we call the user's pathway.

When buying music online or ordering food, the stakes are relatively low. When a public system experience goes wrong, it can have real consequences for a person, a family, a business or an intermediary. This could mean anything from a person choosing not to participate any further in something that would have benefited them or their community, right through to a serious impact on an individual's or community's quality of life. Whatever the consequences, design approaches take human experiences into account when seeking to understand a system.

In fact, design thinking makes people's experiences as important as the products or services being designed. It turns experiences into scaffolds around which products or services can be constructed. Using a design approach makes our process of problem solving different – we elevate the user experience as an entry point for designing a way to improve it.

The service, product and channel layer

This layer looks at the services, products and channels of interaction that make the user experience.

Some design academics argue that experiences can't be designed because an experience is dependent not only on what is provided but on the background of the service recipient. Two people experiencing exactly the same product or service can have a very different view of it, as shown by feedback websites where the same product or service receives a highly critical one-star rating from one user and a highly affirmative five-star rating from another. A look at any hotel, attraction, airline or restaurant will show a similar spectrum of experiences.

However, the apparent tension between these results does not mean that designing experiences is a useless endeavour. Websites that rely on user feedback often calculate the mean of the results to give a better understanding of the overall spread of responses. With some refinement of results, the data shows that overall some products and services do fare better than others in the consumer ratings stakes. Rather than disempowering designers, this example shows that design should not intend to create identical experiences for all users of any given service.

After all, people don't experience a system as such. They experience a pathway through a system, a pathway that is shaped by the numerous products and services that form that experience, and any organisation (or collection of organisations) is unlikely to control all elements of the experience. For example, in a health system interaction the experience is shaped by the individual medical professionals and allied health professionals that a person deals with, and the way the system collects and manages information. Some of these will be self-employed and others will be employed by an organisation, such as a national or state government health department, a non-government organisation or a corporation. These organisations shape aspects of the health system, including overall frameworks, treatment services and diagnostic services.

The delivery layer

The underlying layer for designing within a complex system is the delivery layer, which is highly detailed. Fundamentally, we are asking questions about how the products and services are delivered.

The model in Figure 9 synthesises the layers of design and shows the delivery layer – the base of the triangle – in more detail. This model is about the levers available to a designer to effect change. In its simplest form we think about people, process and technology, but this model is more sophisticated and puts more levers at the designer's disposal. These levers are important to think through when managing change and designing an organisation.

In exploring the delivery layer, focus on these key questions:

- Networks – What soft relationships can be put in place to improve the design?

- Systems and technology – What technologies will deliver the experience? Are there opportunities to harness new mobile technologies?

- Processes – What new or changed business processes will be required to bring about the change? Can the business processes be streamlined and made easier for those who experience them?

- People – What workforce is required for the change? What skills, experience, numbers, levels, locations and culture are required?

- Structures – What structural changes can be made? How can resources be better allocated? What policies can be changed?

Figure 9: Elements of design layers

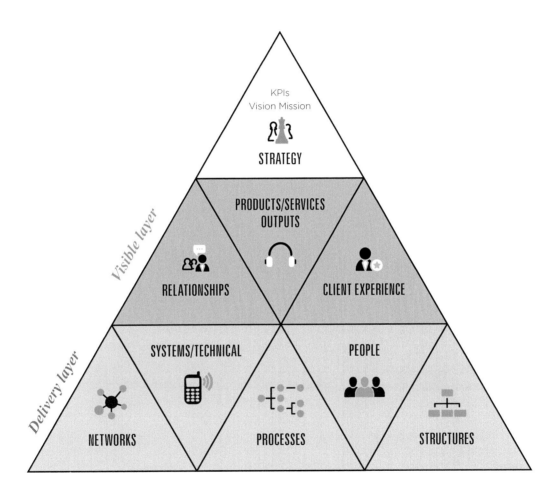

Turning vision into reality

The four layers of design are a necessary and useful starting point but insufficient to effect change. Effective change results from implementation of the intervention, which includes project management to bring the various products and services into reality, and working with people so they understand why the change is valuable, what the change is and how they need to respond.

A core function of the designer is equipping people with the tools and techniques that help them to not only manage change but also to effectively lead individuals and organisations through times of change.

This process does not start when the change is ready for release – it starts on day one of the project because it is intrinsic to the co-design approach. This is one of the great advantages of co-design. The core design team, by virtue of its close engagement throughout the project, has been prepared to implement the change. And the co-design process, by engaging the people who will be most affected, increases their acceptance of the change and expands their capacity to participate in it.

There are three key types of change:

- Developmental change – improving the performance of an existing system.

- Transitional change – transitioning to new processes or technological platforms while maintaining the inherent structure of an existing system. This may reconfigure the system but its intent and purpose is fundamentally the same.

- Transformational change – a change to the underlying strategic intent of a system. Transformational change of a system becomes necessary when there is a seismic shift in the underlying paradigm and the existing system ceases to be viable in the face of anticipated change.

The Design System enables transformational change in complex systems by:

- applying a human-centred approach

- establishing design leadership

- building a shared vision for change through co-design.

A human-centred approach to change

By articulating and focusing on the desired human experience, the designer in a complex system can provide a catalyst for change. They move organisational and systemic thinking beyond the architecture of the current state as it is defined by existing policy and process. This fresh, forward-thinking, experience-based vision of the future creates a space of possibility and defines new outcomes to guide a new intent.

Establishing design leadership for change

Effective change relies on creating engaged leaders who understand the need for change from different perspectives. These leaders build a shared vision and sponsorship in all the key sectors impacted by the change, creating a sense of cohesion across the system. Change leadership also allows for exploratory research to improve understanding of the problem, the areas where the system is failing to meet current needs, and the impact of future needs and change. Change leaders identify the critical shifts necessary to make the system viable, resilient and vital into the future.

Building a shared vision for change

A co-design ideation process generates multiple options and tests their viability, asking what is desirable, viable and possible from different perspectives. This approach harnesses diverse knowledge and experience, and has the added benefit of using a process of authentic engagement, which creates a shared vision for the way forward and builds trust and accountability with key stakeholders.

GENERATING PUBLIC VALUE

If design is about devising 'courses of action aimed at changing existing situations into preferred ones' (Simon 1988), the designer needs a firm grasp of what 'preferred' means. In complex systems we think 'preferred' is about public value.

We start by understanding public value through the program logic model (Figure 10). Designs transform inputs into outputs, which are then shared with the community to generate greater public value.

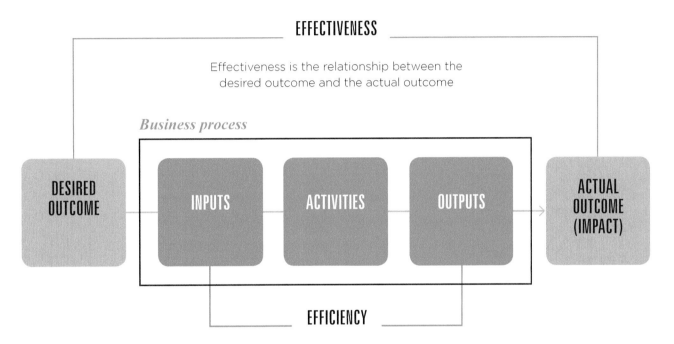

Figure 10: Program logic

More specifically, we can think of public value as a three-stage development: creating, adding and sharing value.

Value is created when an organisation's tangible and intangible assets are combined, through their unique position in society, to produce something. For example, a police force has budgets, skilled people, policies and procedures, as well as relationships with the community and other organisations. These tangible and intangible assets are combined into capabilities that are then able to respond to or prevent crime. These capabilities could include forensics, tactical response or high-tech crime intelligence.

Value is added when capabilities are deployed. For the police organisation, capabilities would be deployed to protect individuals or organisations from violence or fraud. The organisation's value would then increase when crime is prevented or resolved.

Value is shared when the collective value adds scale to others and the overall community benefits. Such shared value is easy to see in our policing example: resolution of crime at a case level has a ripple effect through the overall community by deterring potential criminals and increasing the sense of safety and wellbeing in the community. Any benefit to the community

has many overarching effects, such as increased confidence in government improving overall governance, reduced cost to individuals and organisations through less theft or lower insurance premiums, less pressure on the health system through reduced cases of violent crime, and even increased business confidence in the nation, with direct flow-ons to the economy.

When thinking about value some areas to consider (Figure 11) are:

- Effectiveness – What is the intended outcome for the organisation or group of organisations? To what extent is that intended outcome being reached?

- Efficiency – What is the ratio of inputs to outputs? Can more outputs be achieved from the same or lower resource inputs?

- Experience – Is the experience for those interacting with the organisation or system good for the people involved, whether customers, clients or staff?

- Ethics – Is the right thing being done?

Figure 11: Public value considerations

ETHICAL

Fair and transparent

The right thing to do.

EXPERIENCE

A good experience for the user

What's good for the user? How can we make our system so that it can deliver this?

EFFICIENCY

Efficient for the system

Efficient for all stakeholders, including for industry, NGOs, individuals and government.

EFFECTIVENESS

The intent of the system gets delivered

For example, the intent could be to collect tax, relieve famine, improve the environment, or distribute income to vulnerable people.

SUMMARY

There are several classes of systems, as described by the Cynefin framework – obvious, complicated, complex and chaotic. The design of complex systems is focused, by definition, on complex systems. These kinds of systems comprise many agents that operate independently, yet are not completely random or impossible to predict. Complex systems, at the individual level, are hard to predict but follow predictable overall trends. They display emergent properties, that is, properties that were not designed into the individual parts but are displayed by the overall system.

Moreover, complex systems present wicked problems, which have no easy solution. Rather than involving only two, three or four dimensions, wicked problems are multidimensional. Understanding these multiple dimensions will help to understand the system. Typically, these dimensions are the vector of breadth (the spatial understanding of stakeholders and nodes in the system), the vector of depth (the deep experiential understanding and empathy with the lived experience) and the vector of time (how the system will change over time, how rapid change will be, and how our intended designs might themselves affect the system over time).

Working with complex systems may seem futile given that most are beyond the grasp of any individual to fully understand. However, Richard Buchanan hypothesised that the theoretical field of design offers a framework for working with wicked problems and multidimensional challenges.

A design approach has three advantages. First, effective design research helps to understand the multidimensionality by gathering different perspectives on the system. Second, a constructive design process creates collaboration by using the voice of design to broker a strategic conversation with the three other voices of intent, experience and expertise. And third, the small and tight-knit core design team effectively drives development of the emerging design.

The combination of these characteristics makes the design approach uniquely suited to working with complex systems. Co-design provides the foundation from which we can uncover potential solutions through innovation, divergent and convergent thinking, and rapid iteration. Moreover, a disciplined and flexible approach that views the system in terms of layers – strategy, experience, service and delivery – allows the designer to change 'existing situations into preferred ones'.

The designer may think their task is finished with the completion of an elegant design, yet in reality their work is just beginning. Bridging vision and reality in a complex system requires the unpacking of multiple layers of detail, and only when all layers are worked through will those affected by the system experience the change. To truly create preferred situations from existing ones, the designer must work through to implementation to see changes become real.

Finally, change within complex systems strives to increase the overall level of public value. As designers, our end goal is to create, add and share value across all stakeholders in the community.

03

METHODOLOGY

The Design System™

CHAPTER 3:
METHODOLOGY

This chapter is about design practice – how we as designers take the steps from initiating a project through to generating ideas and realising a shared value result.

Our design practice is underpinned by principles that guide us in the design and execution of every project, without compromise. These principles are, in turn, enabled by a clear design methodology – the Design System™. If the core concepts described in Chapter 2 guide our thinking, the design principles and methodology described in this chapter guide our actions.

The methodology can be framed and adapted according to the particular type of design challenge (as explained in Chapter 4). The design process is therefore not a formula but a unique set of practices.

This chapter is organised into two sections:

- **Design principles** discusses the core principles that guide our projects. These principles ensure a design approach is taken that creates a clear proposition for every design challenge.

- **The Design System** is the core methodology. The system covers four stages: envision, design, make and measure.

"THE JOURNEY OF THE INNOVATOR...IS LEARNING HOW TO 'CUT CUBES OUT OF CLOUDS'. HOW CAN YOU GIVE SHARP EDGES TO A SOFT CONCEPT SO EVERYONE CAN SEE IT? HOW CAN YOU MAKE THE INTANGIBLE TANGIBLE?"

MARTY NEUMEIER
AUTHOR AND SPEAKER
2009
REPRINTED BY PERMISSION OF
PEARSON EDUCATION, INC., NEW YORK

DESIGN PRINCIPLES

A design principle is a statement that guides the designer's actions. It should be timeless and authentic. We've developed a set of design principles that we genuinely embrace and apply in every aspect of our work.

establish a clear intent

Seek focus and direction from the start. Clearly define every project by establishing a clear argument – ask, what is the nature of the design problem or opportunity? This helps to imagine the future by framing a compelling argument for change. At each decision-making stage, maintain integrity to the core rationale for the project.

take a human-centred approach

Work with complexity from the user's perspective. Create change from the outside-in, with people and for people. Put the human at the centre of the design process, starting with a deep understanding of the human experience, and referencing an ecosystem of different experiences throughout the entire process.

drive collaboration and conversation

Harness innovative thinking from multiple disciplines by encouraging people from different perspectives to talk and work together. This diversity of views and experiences helps designers explore and collaborate.

seek exploration and innovation

Foster divergent thinking by exploring from the outside-in and aiming to propel knowledge generation. Drive convergence through directed synthesis towards new possibilities and options, targeting not just improvement but transformation.

visualise and prototype early

Make ideas visible early by iterating concepts and experimenting with new design solutions. As opportunities emerge throughout the process, continue to improvise and maintain a clear line of sight between vision and reality.

seek a balance of desirable, possible and viable	Find the intersection between the user experience (what is desirable?), technical feasibility (what is possible?) and business viability (what is viable?). Seek optimisation where the balance is struck.
follow a disciplined and flexible process	Customise the design process to make it context-specific, approaching each new method and design activity with a fresh perspective. Be rigorous in your thinking and apply methods that you know work, continuing to innovate and improve your thinking.
design the whole system	Take a comprehensive system view through each stage of your work. Uncover the ecosystem of relationships, interactions, dependencies, and unintended consequences. To design working solutions, align individual parts with the whole and design solutions that work.

THE DESIGN SYSTEM

The Design System™ is the methodology we have pioneered based on thousands of projects dealing with complex challenges. This methodology has evolved over decades of application in Australia, New Zealand and Singapore, and in developing contexts such as Kenya. Although we call it a 'system', it's not a prescriptive set of actions. In fact, it's a design methodology that can be applied in a multitude of ways, a methodology we expect will continue to evolve.

Prior to explaining the methodology, an important initial condition is the nature of the challenge as presented by the client or project sponsor. In many cases the problem artificially appears to be defined; what the design methodology implores of the designer is a willingness to frame the project in terms of discovery and redefining of the problem. In public value contexts there are patterns of design challenges (which we explore in more detail in Chapter 4).

The Design System contains many layers of thinking with a bias towards action and co-design. The design process is not always linear, and there are periods of divergent thinking followed by periods of convergence to arrive at a design. The designer in complex systems must be able to navigate towards what should be done next. It's useful to think about the process as interconnected zones of work, rather than strictly consecutive steps.

The Design System is expressed as four diamonds (Figure 12) that encompass four zones of design: envision, design, make and measure. Each diamond incorporates five phases that comprise the typical work necessary to deliver that part of the design process: intent, explore, innovate, evaluate and formulate/implement/recommend.

Envision it – this is the process of engaging strategic imagination about the future, devoid of constraint or concern about how to get there. 'Envision it' explores scenarios of the future, imagines preferred futures and develops compelling pictures. This part of design aims to excite, to evoke emotional responses from the diverse users and other actors about the possible future that they can shape and create. The result is visually compelling futures that are documented, filmed or otherwise represented. These artefacts hold the inspiration for change and galvanise support for taking action toward this preferred future. The future is not a place you are going but a place you are creating.

Design it – is the creation of conceptual frameworks, strategies, policies, services or product interactions, and organisational designs. It's about deep understanding of needs and opportunities to make a change. It involves experimentation to learn and evolve good ideas into propositions for scaling. In complex systems this is a critical step to start small, understand how an initial population might experience a change and co-design the iterations to learn and apply for scaling to more people. The result of this diamond is largely conceptual and abstract because it does not necessarily result in a tangible change in people's day-to-day experience. Instead, we use conceptual design to understand what, how, when, and to whom change may occur.

Make it – is the detailed design and implementation of the solution we conceptually developed in the 'Design it' diamond. Here, we design the change in greater detail, generate prototypes, and transition to implementation of the successful change. We create an enabling environment for the change, ensuring that our design is embedded in practice. We typically dive into detailed product design and implementation by drawing on expertise in business design, process design, people, systems and roles, IT tools, digital solutions and change management. The result of this diamond

is a changed experience for the users and other actors in the ecosystem we have designed.

Measure it – qualifies and quantifies the actual impact or effect of our designs on the initial problem. This diamond focuses on measuring the impact, experience and delivery of the services by an organisation (or organisations). We analyse the extent to which our work has increased public value. This is a critical moment in our design process, because it's where we offer evidence for changing our original conceptual designs and for changing the existing products, services and organisational delivery mechanisms. This diamond generates feedback loops that drive corrective action at the policy, strategy, service or delivery layers.

Figure 12: Design System core methodology

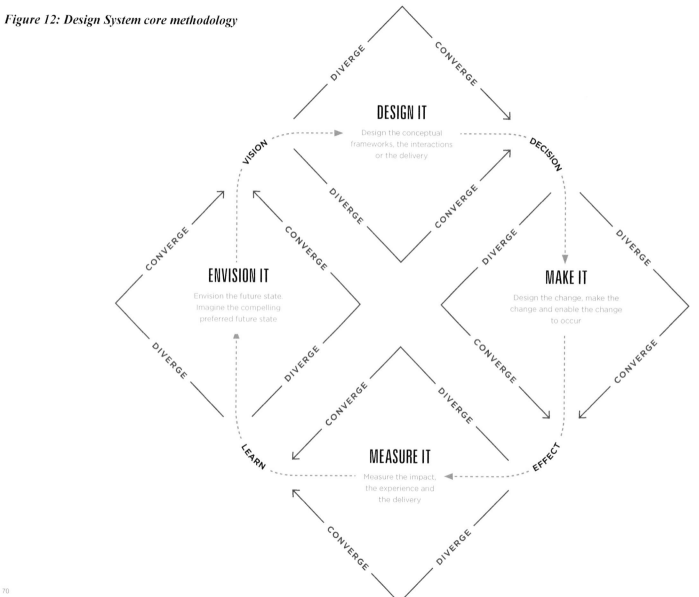

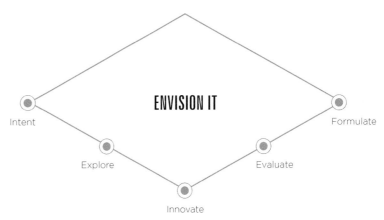

ENVISION IT

Intent

Explore

Innovate

Evaluate

Formulate

ENVISION THE FUTURE
STATE. IMAGINE THE
COMPELLING PREFERRED
FUTURE STATE

Envision it

"What makes us different is that we're brave enough to say that where we don't understand the context, we'll go talk to people directly and learn about it."

Raffaella Recupero
ThinkPlace Australia

The end result of this diamond is an artefact that illustrates, in a compelling and tangible way, the future that has been imagined by the collective voices in the system.

Intent	*Explore*	*Innovate*	*Evaluate*	*Formulate*
Collaborate with key voices in the system and invite new voices to initiate the argument for designing a compelling future.	Research and discover imagined possible futures by inviting provocations and scenario exploration.	Imagine and model possible futures, generating engagement with voices in the system.	Identify preferred futures and underpinning assumptions to realise this future.	Create a compelling and exciting visualisation of the future.

Design it

"There is a real transformation in understanding when the design team starts talking about the solution from the user's experience."

Cate Shaw
Executive Design Manager,
ThinkPlace Australia

The end result of this diamond is a blueprint or similar artefact that captures the insights and ideas for reaching a desired end state that aligns with the core design challenge intent. This stage equips clients with the arguments to take a prototype or idea and scale for impact.

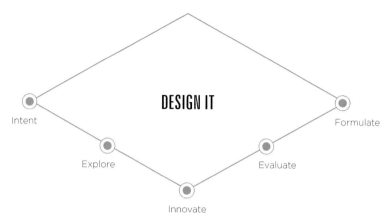

DESIGN IT

Intent

Explore

Innovate

Evaluate

Formulate

DESIGN THE CONCEPTUAL FRAMEWORKS, THE INTERACTIONS OR THE DELIVERY

Intent	*Explore*	*Innovate*	*Evaluate*	*Formulate*
Collaborate with the key voices in the design challenge to define the strategic question for the design project. What is the driving argument for change or action?	Research and discover the human experience of the problem or opportunity. Synthesise insights to drive innovative ideas.	Generate and rapidly prototype new ideas following co-design methods.	Evaluate these ideas with users to refine concepts.	Create a defined proposition that describes the future state and how to achieve it with the core concepts.

Make it

> "We start with a principle of economy – what's the minimum viable prototype that can bring the idea to life? We also keep this process collaborative and get people involved. Test, test, test."
>
> *Leslie Tergas*
> Design Director, Partner, ThinkPlace
> New Zealand

The end result of this diamond is the implementation of solutions whose full effect catalyses positive change.

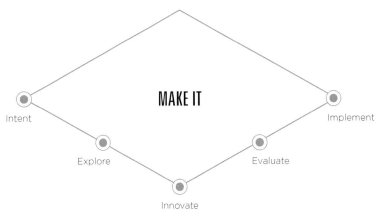

MAKE IT

Intent

Explore

Innovate

Evaluate

Implement

DESIGN THE CHANGE, MAKE THE CHANGE AND ENABLE THE CHANGE TO OCCUR

Intent	*Explore*	*Innovate*	*Evaluate*	*Implement*
Collaboratively engage the voices of design and define the need for implementation.	Discover the human factors and needs that inform the detailed design and structure of the solution.	Generate agile new ideas and prototypes to build solutions. A range of products can be made at this stage: websites, IT applications, new business processes and procedures, new organisational roles and working arrangements, and so on.	Maintain active user engagement throughout the project, ensuring solutions are built to meet user and business needs. Use agile methods to evaluate iteratively.	Embed the change by defining impact and business benefits, communications, co-design training and skill development, and preparing users to transition from the current experience to the new experience using change management strategies.

73

Measure it

> "When evaluating, we hold disruptor activities. We're critical about the proposed design and enlist our colleagues and other interested stakeholders to help us 'break' it."
>
> *John Body*
> Founding Partner, ThinkPlace Global

The end result of this diamond is a measurement framework ready to be implemented. The framework defines the intended outcome and provides a suite of indicators that show whether the implemented change is performing successfully and creating public value.

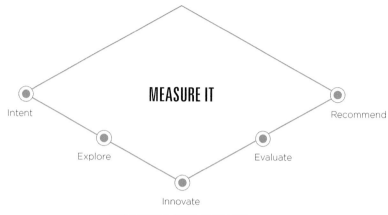

MEASURE IT

Intent

Explore

Innovate

Evaluate

Recommend

MEASURE THE IMPACT,
THE EXPERIENCE AND
THE DELIVERY

Intent	*Explore*	*Innovate*	*Evaluate*	*Recommend*
Collaboratively engage the voices of design and define what needs to be measured. The intent is based on the intended policy, strategy or program intent, and varies depending on the scope of the measurement.	Examine the organisational context and performance, looking at strategic documents. Conduct research with users in the system to frame measurement.	Build different measurement frameworks on a hierarchy of indicators cascading from the highest level of strategic intent. Ensure that the indicators are attributable to the organisation's actions.	Test alternative measurement frameworks to determine which frameworks most closely achieve the intent. Test with different stakeholders to determine which frameworks best describe performance.	Synthesise the measurement frameworks. In addition to converging on the best cascade of indicators, look at practical measurement considerations such as the unit of measure and the source of the information.

CONCLUSION

Our underpinning core concepts locate our practice in philosophical foundations – our way of thinking – especially co-design, and driving innovation through action and learning.

This chapter introduced the core principles that guide our everyday actions as designers. We are driven by our principles and have a bias toward action. The design diamonds of our methodology illustrate the different zones and phases of designing.

The next chapter, 'Design challenges', shows what this methodology looks like in action. We explain the six different types of design challenge we use to model our Design System for specific client challenges.

Designing intranets Creating sites that work

TEIN the life of a GENUS

LEADERSHIP

Six Steps to Transforming Performance at Work

DAVID

APGAR

RISK INTELLIGENCE

Myth Mistery

MICHAEL E. GERBER

the 8th HABIT

04

DESIGN CHALLENGES

*Tailoring the methodology to meet
each client's unique needs*

CHAPTER 4:
DESIGN CHALLENGES

The Design System methodology explained in Chapter 3 is best applied to projects by framing them as specific types of design challenge.

As designers we embrace the most challenging and complex problems because we are driven to make a difference. The key to making a difference is paying attention to the nature of a problem, its unique characteristics, and the issues it raises. No matter the situation, we focus on creating value for the client and for the broader public.

A design challenge has a focusing question or proposition that addresses specific, explicit needs. For example, a service design challenge we worked on in a government context aimed to increase participation in the workforce, especially among high-risk groups such as teenage parents. In this project we defined the design challenge as follows:

Explore how to get teenage parents to participate in education, employment and training for themselves, as well as engage in early childhood development for their children. Specifically:

- What are the preventers to teenage parent participation?

- What are the preventers to childhood development opportunities?

- What are the opportunities or incentives for teenage parent participation?

- What are the opportunities or incentives for childhood development?

By viewing your project as a type of design challenge, you open your mind to the types of questions your clients will likely have and the designs you need to make for them. Design challenges frame why and to what end you apply the design process, tailoring your approach to optimise the desired impact, experience and delivery.

In this chapter we describe six types of design challenge (Figure 13): policy, strategy, program, service, product and organisation. For each challenge we begin by exploring typical client questions, because these are important navigational cues that help us propose the right application of the Design System methodology. Then we discuss the purpose and goals of each challenge, translating the challenge into the language of the Design System – envision it, design it, make it and measure it.

While each design challenge has its own particular goals and characteristics, they don't exist in isolation. Often you will 'stack' different design challenges to apply the design process from vision through to reality. We've provided case studies to demonstrate stacking.

"EVENTUALLY, EVERYTHING CONNECTS – PEOPLE, IDEAS, OBJECTS... THE QUALITY OF THE CONNECTIONS IS THE KEY TO QUALITY PER SE."

CHARLES EAMES
ARCHITECT AND INDUSTRIAL DESIGNER
ATTR. 1998

Figure 13: The six design challenges

1	*Policy*	This design challenge is usually triggered by government. It addresses a social, economic or environmental challenge or opportunity.
2	*Strategy*	This design challenge is usually triggered by an organisation or group. It addresses their vision and strategic intent, and the objectives and strategies to move towards that vision.
3	*Program*	This design challenge is to develop a program of work that needs to be delivered to achieve a policy or strategy aim. It defines the set of projects that would most effectively achieve the desired future.
4	*Service*	This design challenge is the interaction layer involving customers and the organisation or group delivering the service. It addresses the experience, efficiency and effectiveness of service delivery, and includes design of the service channels that people interact with.
5	*Product*	This design challenge is a specific interaction point in a service experience. The interaction might be non-human (e.g. an app, website or paper product), or human (e.g. an advice line or face-to-face interview).
6	*Organisation*	This design challenge is how an organisation or group of organisations functions in terms of its people, processes, tools, information, policies and governance. It addresses the organisational arrangements that deliver services and products to customers.

POLICY DESIGN CHALLENGE

What is the client need?

The types of questions and issues that clients pose if they have a policy design challenge include:

- How can we make the system of government work better for the most vulnerable groups in our communities?

- We have tried so many interventions and haven't been successful. What's a better intervention to achieve our aim?

- What are some ideas we can take to our Minister to bring her attention to the idea that we need to do things differently?

- We have identified limits in our current legislative context. We need to review these limits and understand what needs to change, from the perspectives of policy, law and administration.

These questions and issues come in all shapes and sizes, and can occur at all levels – from multi-country, to whole of country, to the regional or local level.

Why policy design?

Leaders in the public sector are increasingly seeking to take a more collaborative approach to developing policy. Collaboration is key to combining diverse views and experiences, grasping the complexity of problems and co-creating possibilities, and to doing so in a constructive way rather than taking lobbying or combative approaches. Collaboration also responds to the community's desire for genuine engagement, to be heard and to feel their voices count. Collaboration mitigates risk and brings together the best minds to drive innovation.

Policy design involves appreciation of the big systems that shape societies, economies and environments, together with a deep appreciation of the day-to-day lived experience of people. To move beyond generalities we must zoom in from the macro to the micro; to gain context we zoom out from the micro to the macro.

The challenges that policy makers face in the policy design stage include using sound engagement techniques with citizens and communities (including professional communities), and managing the expectations of those communities. Differing stakeholder needs create opportunities for innovation. Where a less experienced person might regard apparently conflicting viewpoints as a problem, a designer experienced in the context of complex challenges sees the opportunity to go beyond a win, loss or compromise.

What is policy design?

The concept of policy design was unpacked and redefined in a major Australian review of business taxation, commonly known as the Ralph Review. The review report noted that:

> The experience in the past has been for policy development, legislative design and administration to be done sequentially with inadequate feedback between the three stages. This has often produced unsatisfactory outcomes from one, or indeed all, of the perspectives involved. (Ralph, Allert & Joss 1999, p. 34)

The review argued for an integrated design process to ensure that policy, legislative, and administrative or compliance concerns are all given appropriate weight and addressed comprehensively in the development of new policy proposals.

Thus policy design can be defined as integration of policy, legislation and administration. When considering policy design we need to engage all voices across the system. This means identifying the policy and legislation owners, and the actors who

are involved in the administration of the law – as well, of course, as the voice of the citizen – to co-design future policies that will work. The designer in complex systems acts as a broker for the different stakeholders, driving constructive engagement among a diverse group of actors.

Goals of policy design

Generate insights: In policy design the end goal may not always be to design a ground-breaking new policy. Rather, the design research may lead to many different insights that result in solutions of all types. Recognise that the demands on policy makers include being more transparent, open, and embedded within community needs and aspirations. The designer is seeking to design through the gap between the policy maker's intent, the citizen's experience and the different areas of expertise of policy administration.

Formulate new policy ideas: This involves working with central agencies and administrative agencies to formulate policy. It's politically driven territory that has contentious boundaries, though increasingly co-design is being embraced for this purpose. Governments ultimately want policy that delivers the desired impact and works for citizens. Design thinking has the capacity to deliver this. Design thinking can be applied to a range of policy contexts to identify and prototype the types of policies government should be putting forward and implementing.

Evaluate existing policy: This involves risk analyses, efficiency and effectiveness reviews, and user-centred research to assess the impact of certain policies in people's lives. It asks: Does existing policy deliver a good experience for citizens? To what extent does the policy achieve its intended purpose? Is the policy efficient for the community and government?

Design system applied

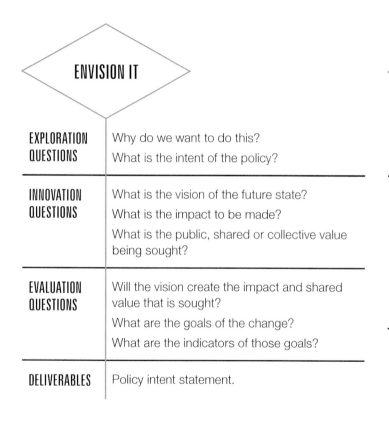

ENVISION IT

EXPLORATION QUESTIONS	Why do we want to do this? What is the intent of the policy?
INNOVATION QUESTIONS	What is the vision of the future state? What is the impact to be made? What is the public, shared or collective value being sought?
EVALUATION QUESTIONS	Will the vision create the impact and shared value that is sought? What are the goals of the change? What are the indicators of those goals?
DELIVERABLES	Policy intent statement.

DESIGN IT

EXPLORATION QUESTIONS	What do we know about the current system? What needs to change?
INNOVATION QUESTIONS	What is the theory of change? What are our big ideas for implementing the policy intent? What do we need to make to implement the big ideas? • e.g. new services, new products, new or changed channels, workforce changes (roles, numbers, skills), IT system changes, apps, websites, transaction systems, communications, marketing
EVALUATION QUESTIONS	Are we confident that the hypothesis for implementing the change will deliver on the change intent? Is it easy to implement and administer? Will this be sustainable with the money and effort we are prepared to dedicate to it?
DELIVERABLES	Policy blueprint.

MAKE IT

EXPLORATION QUESTIONS	Have we used an agile design process for each element that has to be made? What is the context we are designing in?
INNOVATION QUESTIONS	What ideas do we have for solutions? How can we prototype the ideas?
EVALUATION QUESTIONS	How can we test our prototypes? What did we learn from the prototype? Are we converging on concepts that are delivering on the change intent, efficient for all parties and deliver a great user experience?
DELIVERABLES	Policy white paper or other document.

MEASURE IT

EXPLORATION QUESTIONS	What are the areas that we want to understand and measure?
INNOVATION QUESTIONS	What will indicate the impact that we are seeking to make? What would indicate unintended consequences?
EVALUATION QUESTIONS	What is the gap between what we thought and what has happened? What can we do to close the gap? Are there any unintended consequences?
DELIVERABLES	Policy evaluation.

STRATEGY DESIGN CHALLENGE

What is the client need?

A leader becomes interested in strategy to ensure the long-term success of their endeavour. Success comes from working with the external context to seize opportunity and mitigate risk. Chief executives, executives and board members ask for help with issues such as:

- we need to hold a strategic conversation as an executive – we don't know how to strategise

- we need to review our corporate plan and want an independent expert to advise us

- we don't have a clear sense of identity or direction

- our board or executive wants to set a clear strategy for the next three years.

Why strategy design?

There are two goals to strategy design – to ensure an organisation is strategically appropriate and to ensure it remains systemically viable. Or, put another way, to ensure the organisation is doing the right things and doing them right.

'Strategically appropriate' means the organisation is striking the right balance between being intentional – that is, pursuing a defined intent – and responding to its context. It means it is applying its effort to the right things, the priority areas that will make a difference. It is choosing the right interventions or strategies, targeting the right markets, developing the right products and making the right investment decisions.

'Systemically viable' means the organisation is operating in a way that will ensure its short-term and long-term continuation. It means it is doing what it is doing in the right way. Most importantly, it is applying its resources in an efficient way, and its workforce, workflows and systems are fit for purpose. It is considering and addressing all forms of risk to the enterprise.

What is strategy design?

Strategy design is a process of learning. The organisation learns by sensing the external and internal world, making sense of that intelligence to generate insights, then designing in response to those insights. It is about being intentional about where the organisation is heading and setting objectives and strategies to get there. The strategy is tested and progressively refined in action. Although this all sounds obvious, in practice there can be two flaws in the process.

The first flaw is sensing and then acting straight away without due consideration. There is no provision to understand what has been sensed – to 'make sense' – and design, therefore the response may not be strategically appropriate. Immediate action may be appropriate in a crisis but generally organisations have time to identify the best courses of action.

The second flaw is sensing and making sense without ever moving to designing and acting. The organisation becomes too reflective to move forward, always finding reasons not to act. This is the product of a culture that is intellectual but also conservative and inert. A certain appetite for risk and uncertainty is required.

An experienced strategy designer can read when it's time to diverge to get more facts and understanding, and when it's time to converge towards designing and acting. Diverging for too long or converging too fast are both problematic for sound strategy.

There are four focus areas when designing strategy.

The first has to do with context. What is the future strategic context and the current operating environment?

The second is about articulating the desired future state. Why does the organisation exist? How does it operate to achieve this (what is its identity)? What does it do (what is its purpose)?

The third is about priorities. What are the few things to focus on that will make a big difference? Few executive teams can prioritise well. There are always so many things to be done but resources must be focused on what matters. Even an organisation as big as Apple, with all its research and development resources, could not simultaneously develop the iPad and the iPhone. It had to prioritise one (in this case, the iPhone) over the other.

Fourth, there must be a bias toward action. What are the specific activities that will support the strategy? How will success be measured? What are the timeframes?

Goals of strategy design

'Plans are nothing; planning is everything.' Depending on which side of the Atlantic Ocean you come from, this is attributed to President Dwight D. Eisenhower (1957) or Prime Minister Winston Churchill.

The point is that the planning process brings its own benefits, such as exploring contingencies, and setting and sharing direction. Make sure the strategic plan is the product of a conversation with the senior team. A good strategic plan communicates the identity of the organisation, the direction it is taking, the priority areas to address to move in that direction and the way that success will be measured. In a complex world it's not possible to tell staff precisely what needs to be done. Instead, the strategic plan can inspire purpose and direction, and give flexibility to staff to work out how to give life to the senior team's strategic direction.

When strategic planning became popular in the 1980s and 1990s it was often left to a junior-level person to sit in a quiet room to craft a corporate plan. It may have looked elegant, but without seeking out diverse inputs and engaging in robust debate these types of plans provided little value.

Strategy should be designed using strategic conversations that:

- **Explore** the strategic context. What are the broad social, economic, political, technological and environmental megatrends that should be considered? What is the business context of the organisation and its products, services, customers and competitors? What is the operational context inside the organisation?

- **Innovate** the strategic response. Identify the strategic options facing the organisation and progressively understand which will be the preferred option.

- **Evaluate** to converge on the preferred response. Through an iterative process arrive at the agreed set of strategies that are believed to be the best.

Design system applied

ENVISION IT

EXPLORATION QUESTIONS	What are the megatrends we are facing? Do the megatrends present us with opportunities?
INNOVATION QUESTIONS	What excites us about the future? Is there a very different place we should be heading? What is a compelling future that we could pursue?
EVALUATION QUESTIONS	When we think about the vision from multiple perspectives does it still make sense? Have we thought through the critical uncertainties facing the organisation? Are we being bold enough given the uncertainties in the future state?
DELIVERABLES	A narrative of the compelling future state.

DESIGN IT

EXPLORATION QUESTIONS	What are the social, economic, political, technological and environmental drivers affecting the organisation? What are the trends in the space where we operate or do business? What is happening inside the organisation? What are the strengths, weaknesses, opportunities and threats in all of this?
INNOVATION QUESTIONS	Who are we as an organisation? What do we do? Why do we do it? Where are we heading? How can we describe this in a way that people can picture? How can we break this bigger picture down into a series of objectives and strategies? What should we start, stop or change?
EVALUATION QUESTIONS	Are our objectives and strategies the best set? Have we tested them in a set of alternative scenario futures? Will our set of strategies make sense in a number of alternative futures?
DELIVERABLES	Outputs of strategic conversations. Environmental scan. Alternative future scenarios. Measurement framework.

MAKE IT

EXPLORATION QUESTIONS	Who is the audience for the strategy?
	What do they need to know about the strategy?
INNOVATION QUESTIONS	What is the best way to involve those who need to understand or contribute to the strategy?
	How can we present our strategy in a way that people will value and understand?
EVALUATION QUESTIONS	Is the strategy fit for purpose?
	Is it understandable?
	Can people engage with it and relate it to their work?
DELIVERABLES	Strategic plan.
	Strategic dialogues.
	Engagement platforms.

MEASURE IT

EXPLORATION QUESTIONS	How will we know if our strategy is working?
	How will we know if the environment has changed to the point that we need to change our strategy?
INNOVATION QUESTIONS	What are the best indicators to show we are performing?
EVALUATION QUESTIONS	Are we performing according to plan?
	Are we conforming to all relevant standards and laws?
	Do we need to recalibrate our strategy?
DELIVERABLES	Strategic review or evaluation.
	Annual report.

PROGRAM DESIGN CHALLENGE

What is the client need?

People facing a program design challenge typically raise issues such as:

- we need to know all the investments we are making and prioritise them

- we can't do everything – even though we want to do it all, we have to decide what is most important, what will make the greatest difference

- we need a clear plan of work that we know will get us where we need to be

- we need to allocate our resources to the right set of initiatives

- we have to demonstrate to the organisation, our stakeholders and government that we have the right mix of investments, initiatives or projects to implement our strategy.

Why program design?

Program design is a critical challenge for organisations looking to manage a suite of change initiatives while continuing to conduct business as usual. Decision-makers have the challenge of effectively balancing business-as-usual activities and projects to improve the efficiency, effectiveness and experience of services. In striking this balance, they want to understand the best way to organise their initiatives to fix, evolve or transform the organisation so that it meets its future vision and goals.

An organisation that is 100% efficient in delivering today's work leaves no capacity for ensuring it will be appropriate and relevant for the future. Therefore being 100% efficient today is a strategy for going out of business, because there is no capacity for innovation, research and development. If services are static they are gradually becoming obsolete.

Program design is about choosing the right mix of investments or projects to implement the strategy. In a fast-changing industry like technology, over half an organisation's resources may be invested in this type of work. Organisations that operate in a slower changing context can invest less in change, but they still need to invest at least 5–10% of their resources in preparing for the future. Less than this level of investment means the organisation is going backwards with respect to its context, peers and competitors.

What is program design?

Program design is the identification and organising of projects that have the intention of improving an organisation's performance. In practice this means that programs are designed to continue the strategic narrative into concrete, achievable projects that will realise the organisation's vision. The emphasis in program design is the creation of outcomes. Decision-makers will have a clear rationale for investing in an array of projects because program design has clearly stated how clusters of projects will come together to ensure a collective impact or end state for the organisation. The point of having a program designed is to exploit economies of scale, reduce duplication of effort across the organisation and increase success for all investments. Program design seeks to identify:

- the strategic drivers and organisational intent

- critical areas of difference that need to be focused on

- where to invest to achieve strategic objectives

- the benefit and impact these investments provide to clients, staff and stakeholders

- the time horizon to implement these investments (one, two, five and ten years)

- sponsors, leaders and deliverers of the initiatives

- the governance required to ensure that the investments are delivered on time and budget, and achieve the intended outcomes.

Goals of program design

Program design is about doing the right projects to deliver the strategic vision. The program designer works with senior decision-makers to design their program of work at the enterprise and division level, helping them frame their programs in terms of importance and benefit to people and the organisation's outcomes. Program design can be proactive or reactive:

- Proactive program design – This might be part of the annual strategic planning cycle that follows the formulation of a strategic plan; or a program of work to implement high-level design recommendations for the change or reform of an organisation.

- Reactive program design – 'In-flight' programs that require a realignment or reorganisation because projects are not delivering collective impact or the context has changed.

This means decision-makers can be in different frames of mind when approaching development of their programs – measured and taking time, or urgent and under pressure.

Choosing how to allocate the investment resources of an organisation is the non-delegable work of the senior team. It requires the right balance of careful analysis and intuition to achieve the optimal investment mix. These choices can be the difference between organisational success and failure.

A program designer must:

- engage the senior team in a series of strategic conversations to ensure the program of work outcomes align with the strategic vision for the organisation

- help the senior team to develop the initiatives requiring investment

- communicate the program such that people in the organisation can implement it.

The sole focus of program design is to design the optimum program of projects to make best use of the investment dollar. That's easy to say but much harder to do, because it involves narrowing an infinite set of investment possibilities to one set.

The most common output of program design is a program of work, an investment plan or a capital expenditure plan.

Design system applied

	ENVISION IT
EXPLORATION QUESTIONS	What are the objectives in the strategic plan? What is the overall program seeking to achieve?
INNOVATION QUESTIONS	What are some of the features of an innovative investment program? How can we have a more exciting set of investments that seizes what is possible, rather than merely seeking to avoid risk?
EVALUATION QUESTIONS	When we think about the program from multiple perspectives does it still make sense? Are the investments locking us into one future, or will they help us to be agile in responding to external changes, whatever they may be?
DELIVERABLES	Program vision. Draft set of investments.

	DESIGN IT
EXPLORATION QUESTIONS	What are the investment options? Are we stretching our thinking about what is possible? What is the total capital expenditure budget?
INNOVATION QUESTIONS	What are some investment mix options? Are we stretching our thinking about what we could stop, start or modify? Are there things that we have always done that we could shrink, grow or reverse? Could we outsource? Or insource?
EVALUATION QUESTIONS	Have we got a clear sense of our criteria for prioritising investment? Should we be looking to investments that save money first? And then reinvesting the savings in big step change? Will the investments deliver on the strategic plan? Do they link to the objectives?
DELIVERABLES	Program design document, including program design outcomes, major projects, milestones, phased benefits realisation, accountable people, and governance.

MAKE IT

EXPLORATION QUESTIONS	Who is the audience for the program? What do they need to know to make the financial decisions?
INNOVATION QUESTIONS	Have we considered sufficient options? Do we have any blind spots? Have we spoken to a diverse range of people?
EVALUATION QUESTIONS	Have we properly assessed the program across the range of investment prioritisation criteria? Is it in a form that the executive or board can endorse?
DELIVERABLES	Processes, tools and reports for a design and program management authority. Other outputs as developed with the client, such as reports, integrated program milestone maps, co-design and communication plans, and program risk management.

MEASURE IT

EXPLORATION QUESTIONS	How will we track our program? Have we covered schedule, budget, quality, functionality and risk?
INNOVATION QUESTIONS	Are we targeting our measurement on the few things that really matter?
EVALUATION QUESTIONS	Is the program delivering on the strategic plan? How will we identify variances? Are program governance arrangements in place? Do we have a team tracking the progress of the investment program and regularly reporting on it?
DELIVERABLES	Program measurement framework including a range of indicators. Program measurement impact report.

Case study: Program design to move customers online

Program design can have a transformational impact for an organisation.

An infrastructure organisation had over 170 projects that all aimed to meet its goal of taking its services online and shifting customers to these online services. The organisation felt this would produce $20 million in savings, but had no clear assurance that it could actually deliver the purported savings (by shifting services online) in order to operate at budget.

We helped the organisation reframe the 170-plus projects as a program of work based on four user segment needs, and reorganised the projects into delivering on one or more of these user needs. The shift in thinking from projects with outputs, to a suite of projects that contribute to a program with clear outcomes in mind, galvanised the organisational executive. It was truly a program design project of transformation that was all about the customer, and the customer's end-to-end experience of the organisation's system.

Our approach was to prioritise segments of customers, conduct in-depth design research with those customer segments, and discover customer needs for online services. We identified services that could be delivered within the prevailing legislative, technical and user-specific constraints, and outlined the initiatives that addressed barriers to shifting customers online. These barriers included legislative, organisational, political, and user or human factors. Ultimately, we reorganised the 170-plus projects into several user outcomes, developed initiatives and the associated time horizons for delivery, and investigated the ways in which the $20 million goal could be achieved.

This project was a testament to a program design that focuses on outcomes, meeting the goals of both the user and the business.

SERVICE DESIGN CHALLENGE

What is the client need?

The types of questions that clients ask if they have a service design challenge include:

- We are receiving a lot of complaints about our service – that it's hard to access, not getting the necessary support – and we want to not only fix these complaints but transform the experience of our customers. How do we do this?

- We need to shift services from paper- and phone-based interactions to online, but our e-services are not getting taken up. Why? And what can we do to increase e-service uptake?

- We need to improve the service experience for our customers but the experience relies on many other services delivered outside of our organisation – how can we do this?

These types of questions are raised by chief executives, executives, program managers and service centre leaders.

Why service design?

Every organisation needs to keep its services relevant, accessible, meaningful and valued by customers. This is true for all types of organisations – private sector, public sector and non-government organisations. User expectations of services are increasing. If an organisation in one sector offers a particular level of service, customers expect that service level to be available across all sectors.

- Customers expect services that keep up with the latest technologies to provide convenience, save time and save money. They want services that are reliable and trustworthy (the latter applies particularly to management of personal information).

- Stakeholders expect services to deliver value within shrinking budgets, demonstrate effective use of shareholder or taxpayer money to deal with complex issues, and keep pace with trends in user-centredness, streamlining, ethical behaviour and reduced costs and burdens.

- Frontline staff expect to have the information, systems, support and ongoing development to meet increasing demands from customers for high-quality service. Frontline staff want to be able to make a difference in people's lives and meet the expectations of stakeholders within service level standards.

What is service design?

Services are designed for people who have needs and goals, and defined by the purpose and role of the organisation delivering the services, whether it is in the private, government or non-government sector. This means service design is a multi-layered design challenge. At its core it is about experiences. These experiences are 'nested' between the organisation's strategic direction and purpose, and its capability to deliver the services. Service design achieves the following:

- Meeting customers' service needs and experience expectations. Service design asks questions such as: how well do customers understand what they need to do? how to interact? what to expect? when they have done the right thing? The service experience is often very complicated, especially in public services and in services delivered by

multiple actors. When we think about a service experience, we need to consider how the service fits into a person's life – for example, it may be a single touchpoint or a repeated interaction. There are four dimensions to frame experience and provide a rich layering of description (Figure 14):

- needs/preferences of the service user

- emotional response the user has to the service

- actions and activities the user must go through with the service

- user understanding and interpretation of the experience.

- Understanding key stakeholders of the organisation and their expectations of service delivery. In the public sector this means government, ministers, the community, staff and other stakeholders. In the private sector it includes shareholders, boards, staff and customers. In service design challenges we must understand these expectations and demands because this is what we are designing for.

- Taking full account of the strategic purpose and direction of the organisation. For example, a private sector organisation can choose which markets it pursues, while a government

service delivery organisation must cater to all relevant market segments in the community. A government regulator must get the balance right between assistance to help people comply with their obligations and enforcement when they don't. Services are the touchpoint of an organisation to the outside world, and the quality of service determines the organisation's reputation. Service design is where the strategic direction of an organisation hits reality.

- Designing an organisation's suite of products and services so that customers perceive a coherent offering.

  - Service design may also design the delivery layer: the core business processes, people, skills, teams, organisational arrangements, and underpinning IT and information required to deliver the service.

The key is to think holistically and systematically about the whole service and service architecture. This is illustrated in Figure 15.

Figure 14: Dimensions in service user experience

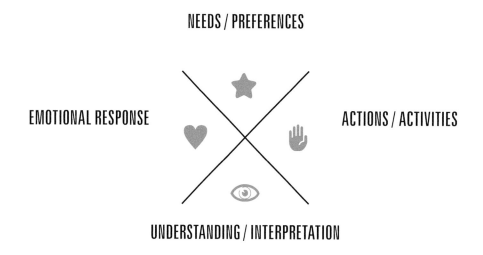

Figure 15: Service design as a business model

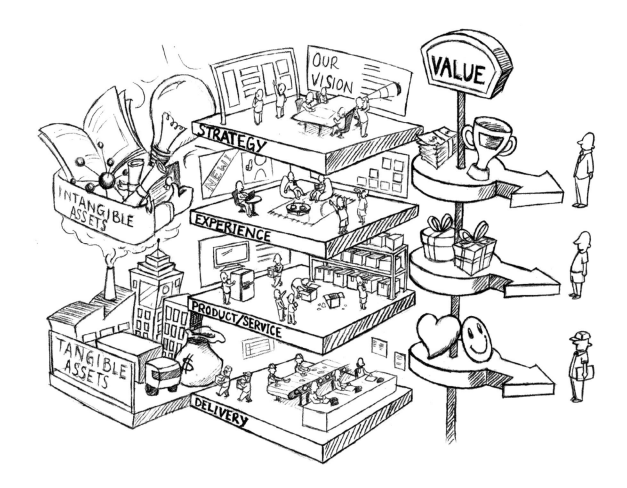

Goals of service design

The art of good service design lies in striking a balance between:

- Desirability – The service should meet the needs and goals of users and the intent of the service provider and other stakeholders.

- Possibility – The service should stretch the bounds of possibility, seeking to leverage new technologies and other opportunities.

- Viability – The service should be viable to deliver, presenting a strong business case.

The scope of a service design project hinges on the capacity, or appetite, for change. Some projects may be focused on marginal changes that help improve the experience of a particular service, while others may aim at larger-scale change that has a more significant impact on the user experience. The appetite for change can be understood according to the following scale:

- Improvement – The focus is the usability of the service – what works, what doesn't, and how people compensate for this. The scope is limited to the existing service and its context of use.

- Evolution – The focus is the usability of a service and its usefulness. The research phase includes investigation of areas of opportunity for extending the existing service.

- Invention – The focus is understanding the user experience. The scope is defined by areas of opportunity for supporting the existing user experience in new ways.

- Transformation – The focus is understanding the user experience and fundamentally transforming the service to better meet user needs.

The goals of service design are based on the appetite for change and include:

- co-design the service with customers, frontline staff and other key actors in the service systems

- identify needs, pain points, and opportunities for improvement, evolution, invention or transformation

- prototype new service models, and iteratively test and validate the designs to ensure an achievable model with clear benefits

- develop a practical pathway of action to implement the new service design.

A typical output of service design is a service blueprint that addresses all four of the layers of service design shown in Figure 15.

Design system applied

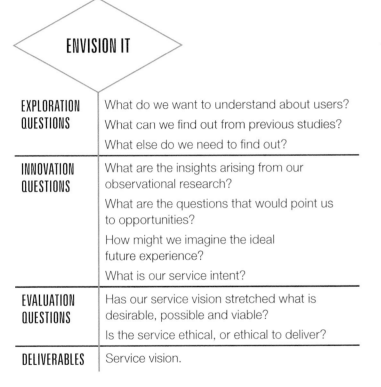

ENVISION IT

EXPLORATION QUESTIONS	What do we want to understand about users? What can we find out from previous studies? What else do we need to find out?
INNOVATION QUESTIONS	What are the insights arising from our observational research? What are the questions that would point us to opportunities? How might we imagine the ideal future experience? What is our service intent?
EVALUATION QUESTIONS	Has our service vision stretched what is desirable, possible and viable? Is the service ethical, or ethical to deliver?
DELIVERABLES	Service vision.

DESIGN IT

EXPLORATION QUESTIONS

What do we need to know to design the service?
Can we access this information:

- from our own information holdings?
- from publicly available ethnographic research?
- from national statistics?
- from other research findings?

Do we need to commission more research?
What is the everyday experience for:

- pain points?
- delight points?
- what works?
- what needs to change?
- where value is being lost/eroded?
- where value could be created?

Important insights are often latent – things that people want or need but don't always tell you. Seek out these insights – they may be framed as observations, statements or questions, e.g. 'how might we…?'

INNOVATION QUESTIONS

What ideas have we generated?

Strategically, what are we seeking to do with this service?

What would delight users?

How might we transform the service experience?

How can we deliver the service...what is the business model?

What can we leverage?

Can we remove or add some steps, or change, grow or shrink steps?

How can we prototype our ideas?

EVALUATION QUESTIONS

How does the service prototype test with users?

Is it easy to use? Is it efficient to deliver? Does the service meet user and organisational needs?

Have we challenged the status quo?

Is the service decision ready to inform an investment decision?

DELIVERABLES

Service blueprint including layers describing service strategy, service experience, products and services, and how the service will be delivered (people, process and technology implications).

MAKE IT

EXPLORATION QUESTIONS	Have we got a service blueprint and implementation plan from the 'Design it' stage?
	What do we know about the delivery context?
	What do we know about workforce skills, numbers, structures, business processes and IT systems that will be affected by the service change?
INNOVATION QUESTIONS	Who needs to be involved in the design of this service?
	What platforms can we reuse to develop this service?
	How can we most efficiently deliver the service while delighting the customer and delivering on the service objective?
	What is the business model surrounding the service?
EVALUATION QUESTIONS	Can we pilot the service before rolling it out more broadly?
	Is the service desirable, possible and viable?
DELIVERABLES	Piloted service ready for widescale implementation.

MEASURE IT

EXPLORATION QUESTIONS	What is the usage of the service?
	Are customers satisfied with the service?
	What does the service cost to deliver?
INNOVATION QUESTIONS	What are the insights from the findings on the research?
	What changes can we make to improve or refine the service?
EVALUATION QUESTIONS	Have the refinements improved the service experience or the efficiency of delivery?
DELIVERABLES	Service measurement framework.
	Measurement and reporting.
	Service evaluation.
	Service impact report.

Steph Mellor, Senior Executive Designer, ThinkPlace Australia

Unpacking the layers of service design

The value we create by doing service design is best demonstrated by looking at the 'Four layers of design' model.

This model is what we use to think holistically about what a business model looks like. We're systemic thinkers, so dividing a business model into layers allows us to consider how a change in one area would cascade into another.

When we think about services, it's about how they fit into someone's life journey. A service may be a single touch point for people, or it may be a repeated interaction. It's important to understand the way the user and the service interact with and inform each other as the user moves through the broader system. The model helps us think about making services which are architecturally sound. That is, we want to make sure we've not only thought about the service in isolation, but we've thought about the people and processes that would need to be in place to support the service.

The **strategy** or **intent layer** is about context, drivers, direction, outcomes, benefits and shifts.

The **experience layer** is about understanding the user journey, motivations and goals, and designing interactions. It thinks about how the experience will look and feel for the user.

The **service layer** describes the specific products, services and channels that the user interacts with. These might include payment services, information products, service charters and channel strategies.

The **delivery layer** is about technology, systems, data, staffing, roles, resourcing and processes.

PRODUCT DESIGN CHALLENGE

What is the client need?

The types of questions and issues that clients pose if they have a product design challenge include:

- Our organisation is changing and we need to communicate to our staff and key clients – can you help us fix the web, and create communication products?

- We have an online service that is failing to meet take-up targets, and we don't know where the problems are, or how to design a solution.

- We need to develop a coherent communication product that explains why we are changing our service.

- We can't afford to continue to deliver our services the way we currently do. How can we shift to a self-service model?

These product questions are raised by program managers, service centre leaders and product owners.

Why product design?

Users expect products and services they interact with in everyday life to meet their needs and be easy to use, accessible and reliable. Digital products are increasingly the only touchpoint that people have with an organisation or system, so their whole perception of the organisation or system is based on its digital products. If they are good, they engender trust and build reputation. If they are poor, they have the opposite effect.

In product design, usability is no longer an add-on – it is a core expectation. Every organisation wants its products to be used and to work for its customers. For government clients, product design must be geared toward the service user, not the legislation or technical perspective, so that people understand what is expected of them and can fulfil the requirements. User-centred products are essential.

What is product design?

The definition of 'product' is broad, and intentionally so because of the broad suite of products that an organisation may offer to deliver on its mission. Products could include apps, websites, online forms, videos, communication products, information products, signage, call centre services and face-to-face counter services.

In addition to these customer-facing products, there are also staff-facing products to help people do their jobs and maintain focus on strategic direction – information tools for staff, intranet sites, reference manuals, communication products, and communication strategies, blueprints and plans. This broad product architecture is shown in Figure 16.

Figure 16: Product design – product architecture for service users and staff

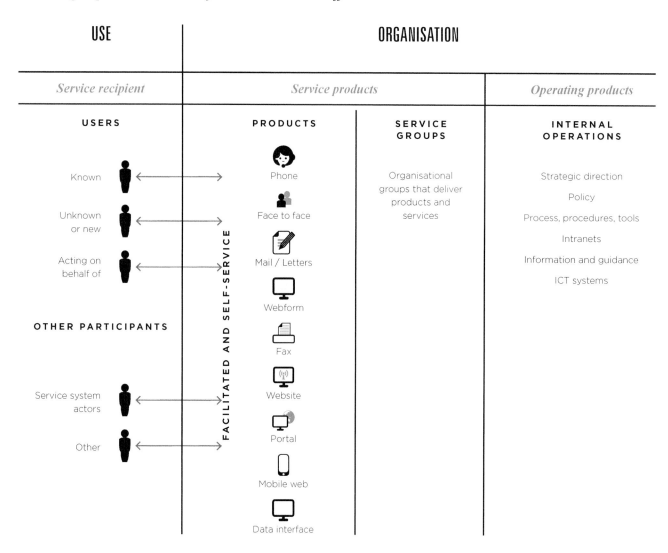

Goals of product design

Product design is unique because different products require very different design expertise – for example, different skill sets are needed for web, print, visual presentation and animated products. Common across all products is the need to understand the user and the situated use of the product. This means understanding humans, the factors that drive their behaviour (such as desire to use the service) and the behaviours that are intended (such as compliance with an obligation). This information is used to build products that are useful, usable and desirable, and meet universal design standards. Product design should take a multidisciplinary approach, working with a range of experts in and beyond the organisation such as IT, design areas, brand and communications, operations, human resources, legal and other corporate enabling areas. The goals of product design are to:

- take a multidisciplinary approach to create products that are not only visually appealing but also highly functional, usable and implementable

- apply user-centred design from concept, through detailed design, build and implementation

- understand the change and transition issues, and develop strategies to ensure product success

- achieve clear benefits from products that allow success to be measured.

Design system applied

ENVISION IT

EXPLORATION QUESTIONS	What information do we want to understand about users? What can we find out from previous studies? What else do we need to find out? What are the scope, budget and expected timeframes for delivery?
INNOVATION QUESTIONS	What are the insights arising from our observational research? What are the questions that would point us to opportunities? How might we imagine the ideal future experience? What is our product intent? The focusing question depends on the type of product and audience – for example, how can we design a product that helps service users access the information they need?

EVALUATION QUESTIONS	Has our product vision stretched what is desirable, possible and viable? Is the product ethical, and ethical to make and deliver?
DELIVERABLES	Product vision.

DESIGN IT

EXPLORATION QUESTIONS

How should we conduct research with users and experts to identify needs?

Has the user research helped us define high-level design or business requirements of the product?

What do we need to know to design the product?

- Does this information already exist somewhere and can we access it?
- Do we need to commission more research?

What is the everyday experience for:

- pain points?
- delight points?
- what works?
- what needs to change?
- where value is being lost/eroded?
- where value could be created?

Important insights are often latent – things that people want or need but don't always tell you. Seek out these insights – they may be framed as observations, statements or questions, e.g. 'how might we…?'

INNOVATION QUESTIONS

Generate product concepts and prototypes to explore how the product might work. These might include low-fidelity paper prototypes of screen designs, or business scenarios of change processes enabled by technology change.

Strategically, what are we seeking to do with this product?

What would delight users?

How might we transform the experience of the product?

How can we deliver the product? What is the business model?

MAKE IT

EVALUATION QUESTIONS

Have we evaluated the proposed product designs against the design criteria that the organisation has defined, including:

- budget?
- timeframes?
- existing products to either leverage or build on?
- build versus buy requirements?

This is a collaborative approach with the core design team, including the voice of intent.

Is the product design decision ready to inform an investment decision?

DELIVERABLES

Product blueprint, including layers describing product strategy, user need and product experience, design of the product and how the product will be delivered (people, process and technology implications).

The product design is typically accompanied by a business case for investment.

EXPLORATION QUESTIONS

Have we got an intent statement, service blueprint and implementation plan from the previous phases?

What do we know about the delivery context?

What do we know about workforce skills, numbers, structures, business processes and IT systems that will be affected by the service change?

The approach in this stage depends on the type of product. It could be digital (a website, app, mobile app), paper (a brochure) or an IT application to enable a business process. It could be new or a refinement of existing products.

Stages include:

- *an intent briefing to determine the scope, context and outcomes of product design*
- *detailed user research to gather the detailed product design requirements, such as:*
  - *information review and pattern scan with key project staff and stakeholders to learn about the limitations and opportunities of the product*
  - *user research with product users (existing or potential) to define the detailed product requirements.*

INNOVATION QUESTIONS

Who needs to be involved in the design of this service?

What platforms can we reuse to develop this service?

How can we most efficiently deliver the service while delighting the customer and delivering on the service objective?

What is the business model surrounding the service?

Iteratively design, build and user test the product. This involves agile approaches to designing and building.

EVALUATION QUESTIONS

Can we pilot the service before rolling it out more broadly?

Does the service optimise desirability, possibility and viability?

This step involves user evaluation of the product build. Testing continues until final acceptance of a product before it goes live. The aim of testing is to simulate the real-world context as closely as possible.

Testing should also cover product implementation and transition.

DELIVERABLES

Piloted products ready for widescale implementation.

Communication products to support the implementation of the new product.

MEASURE IT

EXPLORATION QUESTIONS	How is the product being used?
	Are clients satisfied with the product?
	What does the product cost to deliver?
	What benefit is the product providing?
INNOVATION QUESTIONS	What are the insights from the research findings?
	What changes can we make to improve or refine the product?
EVALUATION QUESTIONS	Have the refinements improved the product experience or the efficiency of delivery?
DELIVERABLES	Product measurement framework.
	Measurement and reporting.
	Product implementation.
	Benefits / impact realisation report.

Mondy Jera, Director, ThinkPlace New Zealand
Rose Wu, Designer, ThinkPlace New Zealand

Case study: Walking the land virtually

We were involved in a digital design project that was really fascinating in terms of how it challenged our assumptions and required a deep amount of social inquiry.

ThinkPlace New Zealand has experience working with Māori, and we recently completed a project related to that population's continued engagement with their land. Our client, a Māori land trustee, came to us not knowing exactly what they wanted, but they knew they needed to re-energise engagement with their landowners, get to know them better and ultimately develop a digital tool that would replace their heavily paper-based correspondence. The organisation had begun to develop a 'Land options report' that they intended to use to maximise the economic potential of land for their Māori landowners.

By way of framing their problem, the client explained that Māori have a longstanding tradition of 'walking the land'. Today, even though there is a great deal of absentee ownership (owners can't be physically present on the land every day), the population still expressed a need to feel connected to their land and 'walk it' however possible. We initially thought that 'walking' the land would be challenging to translate into an online experience, but expected our research would illuminate some of the ways people would relate to this kind of digital service offering.

We interviewed 15 Māori landowners to explore what was important about being a landowner, their aspirations for their land, their relationship with the trustee and what improved communication would look like.

We found that while economic maximisation was one aspiration of landowners, it was perhaps not the most important goal; protecting Waahi Tapu (sacred land), ensuring their children and grandchildren had a place to return to for spiritual regeneration and support, a place to build a house or even to work and live off the land were among the other goals cited by our participants.

We prototyped a digital engagement tool that more closely reflected the customers' needs by aiding decisions and aspirations for their land, such as:

- maps with location and land area

- the number of other landowners

- the related lineage, or whakapapa

- the lease information

- the current use and income generated from the land.

The result was that people who could not actually be on their land could feel more connected to it through locational maps, and also feel more in touch with the land trustee.

ORGANISATION DESIGN CHALLENGE

What is the client need?

The types of questions and issues that clients pose if they have an organisation design challenge include:

- Our organisation is undergoing change and we need to review the way we undertake our work.

- I've recently taken a new leadership position and I want to review the division and look for innovation opportunities – I want to shake things up!

- We're experiencing increased pressures to perform, and our current resources are not expanding. We need to work out what to do.

- We have to improve the way we deliver services to customers – we need to look at how we are organised to meet new service expectations.

These strategy questions are raised by chief executives, executives and board members from all types of organisations.

Why organisation design?

Every organisation exists in a dynamic context of changing competitors, changing customer expectations, changing legislation and changing technologies. Executives and leaders of organisations are continually forced to rethink their strategies and the way they operate. There are three common triggers:

- changes to the environment, including changes of government, policy or legislation, worsening customer experiences, and changing stakeholder expectations

- changes to the strategy of the organisation, including its purpose, vision, core functions or mandate

- changes to business operations, including accountabilities, resources, people, processes, technology growth and business evolution.

It's rare for an executive or leader to think their organisation is perfect. There are always emerging threats and opportunities to respond to. The challenge they commonly face is that they feel their responses to these threats and opportunities are piecemeal, rather than addressing overall organisational design. Good leaders recognise that a piecemeal approach leads to sub-optimal organisational performance because of gaps and overlaps in roles and functions (Corkindale 2011). The result is information not shared, gaps in the customer experience and low innovation because context is not shared.

What is organisation design?

Organisation design is about aligning the elements of an organisation – including its people, processes, services, relationships and structure – with its outcome and purpose. Good organisation design considers all elements holistically, understands how they work independently and together, and rebalances them to bring about an overall improvement in efficiency, effectiveness, and the satisfaction of customers and stakeholders. Most people think of organisation design in terms of structure, but there are many other elements that are equally important.

Like the human body, an organisation has various systems that begin at a micro level and come together to form the complete whole. In a healthy organisation, the different parts should work together. The 'anatomy' of an organisation is made up of:

- strategy – its identity, purpose and direction, including the markets or sectors in which it operates and the customers it serves

- functions – its work, jobs and accountabilities

- processes – how the work is executed

- structure – how it bundles its work, jobs and accountabilities to enable the strategy

- capabilities – the expertise and capacity needed to deliver the work and functions, including people, skills, processes, tools, technologies and information

- culture – the values and behaviours that define its 'personality'.

Organisation design starts with a clear intent that identifies the challenges and opportunities to be considered. This is followed by a high-level design to ensure that the intent and drivers for change are well understood, and that the change that needs to occur is grounded in experiential data and facts.

Goals of organisation design

Organisation design is a conscious and deliberate effort to design (or redesign) an organisation for the context and time within which the organisation operates – to achieve a desired outcome within a set of conditions. It is the alignment of all the elements of an organisation, considering all the elements holistically, understanding how they work independently and together, and rebalancing them to bring about some form of overall improvement. Ideally the improvement would be in efficiency, effectiveness and the satisfaction of clients and stakeholders. The goals of organisation design are to:

- understand the organisation's goals and operating context, and define the required organisational design to meet these goals and operate effectively

- define the desired experience of customers and staff to ensure the design of the organisation supports these experiences

- define the business model identifying the core functions of the organisation

- define the underlying core processes and people required to deliver these functions

- define the work teams, skills and competencies to deliver the processes

- define the enabling capabilities such as IT applications, data and knowledge

- craft a compelling organisational design narrative to drive the change.

Design system applied

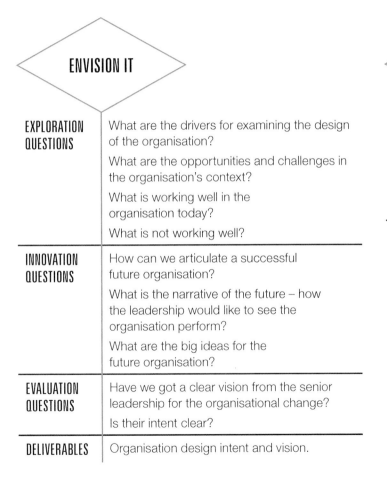

ENVISION IT

EXPLORATION QUESTIONS	What are the drivers for examining the design of the organisation?
	What are the opportunities and challenges in the organisation's context?
	What is working well in the organisation today?
	What is not working well?
INNOVATION QUESTIONS	How can we articulate a successful future organisation?
	What is the narrative of the future – how the leadership would like to see the organisation perform?
	What are the big ideas for the future organisation?
EVALUATION QUESTIONS	Have we got a clear vision from the senior leadership for the organisational change?
	Is their intent clear?
DELIVERABLES	Organisation design intent and vision.

DESIGN IT

EXPLORATION QUESTIONS	What is the senior leadership intent for organisational change described in 'Envision it'?
	What research do we need to conduct about the organisational environment including clients, staff and current practices?
INNOVATION QUESTIONS	What options can we generate to meet the organisational design intent?
	Generate innovation options to redesign or design the organisation. This is driven by good organisation design criteria, or 'organising principles' – which optimise the desired outcomes.
	By having agreed criteria and then co-designing options, much of the anxiety of organisation design is reduced.

MAKE IT

EVALUATION QUESTIONS

How do the design options rate against the organisational design criteria?

Prototype the options with the executive and leaders, the staff and key stakeholders. The process of iteration and refining the prototypes is critical.

Evaluating high-level options often triggers detailed implementation questions. These should be gathered and recorded to be presented in the high-level design organisation blueprint.

DELIVERABLES

Organisational design blueprint – the consolidation of the research, options and evaluation into a compelling high-level design.

EXPLORATION

The 'Make it' zone of an organisation design challenge may involve multiple projects, or a program of projects. Conduct detailed user research to explore the organisational element design requirements (which may involve product design), such as:

- *information review and pattern scan with key project staff and stakeholders to learn about limitations and opportunities*
- *research with users and potential users to define and iteratively develop the product and requirements together.*

INNOVATION QUESTIONS

Have we considered:

- new capabilities, such as new business practice using new IT tools?
- new business processes, and supporting procedures and rules?
- new applications and tools, such as new IT applications and tools?
- new structures, roles, job descriptions and other working agreements?
- new brand and identity for organisations (this impacts leadership, product design and new business processes and practices)?
- business implementation and transition? (this step does not wait till last, but rather is designed well before implementation)

EVALUATION QUESTIONS

Evaluate the detailed elements being designed and prototyped against the evaluation criteria:

- is the client experience being delivered?
- how well does the design deliver on the organisation's mission?
- how efficient is delivery?
- what is the staff experience?
- does the design support overall integration of the organisation?

DELIVERABLES

Prototypes that are tested and ready for implementation of:

- strategic direction
- workforce
- skills
- functions
- business processes
- systems
- resources to support staff.

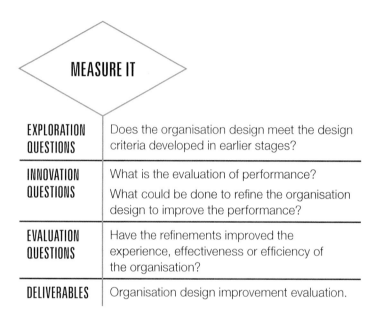

MEASURE IT	
EXPLORATION QUESTIONS	Does the organisation design meet the design criteria developed in earlier stages?
INNOVATION QUESTIONS	What is the evaluation of performance? What could be done to refine the organisation design to improve the performance?
EVALUATION QUESTIONS	Have the refinements improved the experience, effectiveness or efficiency of the organisation?
DELIVERABLES	Organisation design improvement evaluation.

Bill Bannear, Managing Director, ThinkPlace Singapore

The sensing organisation

To work in organisation design you have to first consider what an organisation is, what they do, and how they engage with the world. For organisations that look at the real world, and intervene (either for harm reduction purposes, like law enforcement, or building public goods or value), they can't know the entirety of the real world. They therefore do what they can to sense the real world, and decide where in it, and how, to intervene and make changes. Sometimes they make the mistake of thinking the model they build (their representation of how the world works) is the real world. For me, it's helpful to map the Kolb experiential learning model onto organisation design challenges (Kolb 1984). Thinking about an organisation as a sensing, living thing gives us insight into how they might act and react upon their assumptions about the real world.

The logic is helpful when you're working on an organisation's intelligence apparatus – how it collects and analyses information about the world in order to make better decisions, strategise, and design and execute effective operations.

The 'sensing organisation' strives to optimise its ability to respond to the real world (an evidence-based or intelligence-led operating model). To make best use of resources, there needs to be a balance of effort between sensing what is going on, analysing the collected information, making good decisions, mounting an effective response, and measuring and learning from the results. For example, there is no point having a comprehensive understanding of a problem if you then have no resources to do anything about it. Likewise, action without good intelligence will struggle to be effective. The organisation needs all nodes to be firing equally, or you get bottlenecks and deficiencies emerging. In my work, I've been asking groups to identify the critical node for investment (for example, are you collecting enough information, or is it your analysis, decision-making or response capacity that limits you?).

The Kolb model is a cycle, whereas in reality, work bounces back and forth between different stages of the cycle. If we revisit the Kolb model and re-interpret it to more accurately reflect how a real organisation works, we represent it as a network rather than a cycle (Figure 17). In this depiction, the organisation features nodes of capability. The collection node represents the sensing apparatus, where data about the real world enters for analysis – this is the point where the organisation starts to assemble its 'mental model' of the real world and decide where to intervene. Representing it as a network is intended to remind those in the organisation that it isn't a one-directional cycle – things bounce back and forth between competencies as needed.

Figure 17: How an organisation senses

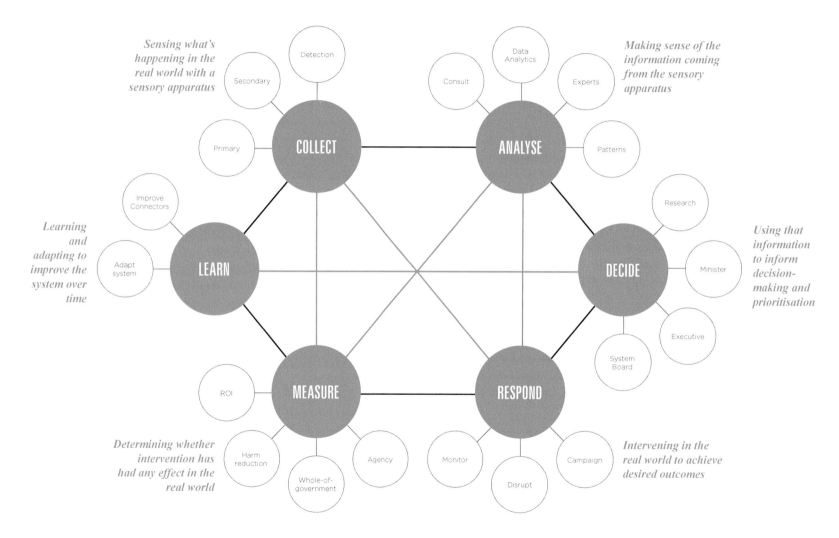

'STACKING' DESIGN CHALLENGES

Our Design System is not meant to be applied slavishly. Our clients' contexts vary and we fashion the Design System to meet their precise needs. Sometimes, this means 'stacking' design challenges – for example, moving from strategy design to program design to policy design. The best way to illustrate this is to review some projects in which we stacked the design challenges.

Case study: Design program of strategic investments + product design + product implementation

We worked with an organisation through three design challenges.

First, we were asked to work with the executive and leadership to define the program of work required to deliver on their strategic intent, or five-year vision. We worked through the program design challenge and developed a strategic investment agenda. This laid out the projects in which the organisation would invest to ensure that it could realise its strategic vision.

Following this work, we moved into defining and resetting the organisation's approach to projects. This opened up the opportunity to develop a new function called the 'Design and Program Management Authority'. This was a rapid organisation design challenge. We designed roles, and tools to be used to enable a new way of running projects.

To help embed the new way of working we ran a series of demonstrator projects, one of which was a new capability to change the way the organisation managed its core work. This triggered a product design challenge, which continued through to implementation of the new capability. The result was one of the most successful business-led IT projects in the organisation's history. This capability is now the core capability for the whole organisation.

Case study: Develop organisation's compliance philosophy + develop tools for staff to follow the new compliance approach

Our client needed to articulate their regulatory approach. We approached this as a strategy design challenge, which involved co-designing the new regulatory philosophy for the organisation and key stakeholders.

To ensure this philosophy was practised, we then worked with the organisation to develop a regulatory handbook for staff. This was a product design challenge that involved co-designing the new regulatory handbook and guidance for staff.

CONCLUSION

Complex design challenges can be categorised into six key challenge types: policy, strategy, program, service, product and organisation. There are hundreds of different design methodologies available but the common elements are:

- having a clear intent of what you are seeking to do
- going through an iterative process of exploring, innovating and evaluating
- converging on and formulating the optimum design.

All design is directed at making a positive difference. When designing in complex systems this means understanding and considering the needs of all stakeholders affected by the change. It also means taking a human-centred design approach while simultaneously considering the whole system.

The design approach to each of the challenges described in this chapter allows those involved to move from ideas through to reality by genuinely co-designing, collaborating and making solutions that work.

Finally, design challenges are not considered in isolation because they are situated in many different contexts (as discussed in Chapter 2). The designer should be not merely cognisant of the context, but have a deep understanding of it – to shepherd the design challenge process towards developing the right design solution. Sometimes this means 'stacking' design challenges to meet the client's unique needs.

05

CORE EXPERTISE

Assembling bold and capable teams to meet each design challenge

CHAPTER 5:
CORE EXPERTISE

In the practice of designing solutions in complex systems, the expertise required is actually a combination of areas of expertise rather than any single expertise. The formulation of design teams should draw together the people who can best explore the design challenge, innovate, and propose changes that realise the design intent.

We have identified 10 core areas of expertise that are necessary to address any complex design challenge. In thousands of design projects we've conducted around the world and across more than a decade, these are the areas of expertise that have proven essential to exploring and solving complex challenges.

In this chapter we describe these core areas of expertise and explain how they may be combined in a single person, or combined in a design team for a particular design challenge.

Additional areas of expertise may be needed for particular projects and challenges, and new expertise is always emerging, such as gamification and innovation labs in recent years. These are noted in this chapter where relevant, but the focus is on the core areas of expertise that apply more generally.

The 10 core design expertise areas that we have built into our practice at ThinkPlace are summarised on the following pages and described in detail later in this chapter.

"NOTHING HAS SUCH POWER TO BROADEN THE MIND AS THE ABILITY TO INVESTIGATE SYSTEMATICALLY AND TRULY ALL THAT COMES UNDER THY OBSERVATION IN LIFE."

MARCUS AURELIUS
PHILOSOPHER AND EMPEROR OF ROME
MEDITATIONS III, 11 (C. 161-180 CE)

Figure 18: Core areas of design expertise

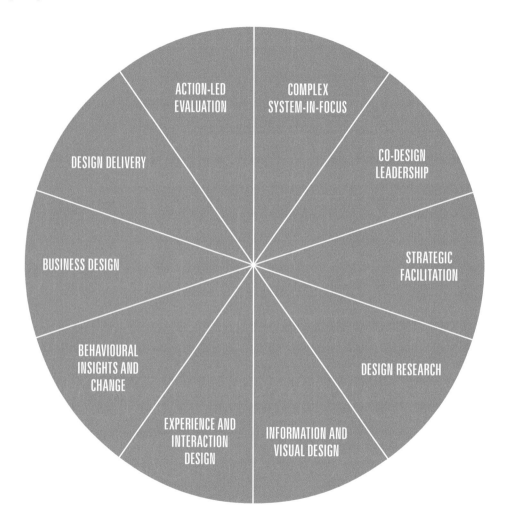

Complex system-in-focus (page 146): A person with this expertise understands the specific domains such as regulation, compliance, development, social cohesion, vulnerable communities and health, and the key concepts, theories and applications of strategies that apply in these domains. This expertise usually comes with many years of experience (at least five) working in the specific sectors and in organisations where real, everyday application has been observed and understood.

Co-design leadership (page 147): A person with this expertise understands the theory, methodology and methods of co-design for a range of design challenges. They are expert at identifying the challenge and then crafting the necessary design program and team. The designers with this expertise have at least three years of experience running projects.

Strategic facilitation (page 148): A person with this expertise can empower, engage, excite, energise and extract best performance from groups of people for design purposes. They are artful in the management of people and information, and have the expert ability to be neutral, broker fresh perspectives and galvanise action towards a common objective. They are highly visual, conceptual thinkers who can grasp concepts and represent them in workshop settings to extract the valuable

wisdom of the group. They break down power structures and give agency for everyone to contribute. A strategic facilitator is also highly expert in strategic thinking and navigation because, unlike other facilitators who are content neutral, good strategic facilitators bring expertise in how to drive strategic change and direction throughout the design process.

Design research (page 149): A person with this expertise is experienced in specific research methods that draw on a range of fields – such as anthropology, social and behavioural sciences, and psychology – and brings strong skills in research design, planning, execution, analysis and reporting. Experts in this area not only apply proven methods but also invent new methods to best discover the human experiences and insights that can inspire design solutions to complex problems. Design research spans three mains goals: discovery, innovation and evaluation.

Information and visual design (page 150): A person with this expertise demonstrates an understanding of information communication principles, engages with data to analyse and seek insight, and produces artful expressions of information in drawn, digital, static, dynamic and interactive forms. The application of visual design to print and digital formats requires a variety of discerning skills. It's critical that visual designers working in complex system design challenges engage with and understand the content and information they are seeking to represent. Merely rendering a good aesthetic is insufficient to communicate complexity and inspire insights for action.

Experience and interaction design (page 151): A person with this expertise develops desired experience models and frameworks that inform the design of interaction touchpoints, such as services and products, with people in complex systems. This person translates design research insights into probable future interactions and draws from emerging technologies and patterns in service systems to create desirable experiences. This expertise operates at both at the macro systems level and the more specific interaction level – for example, the design of digital product interactions.

Behavioural insights and change (page 152): A person with this expertise understands the human behavioural factors driving actions, attitudes and responses to certain stimuli. They draw from behavioural economics, anthropology, sociology and psychology. This expertise contributes to the formation of design research insights and the translation of insights into design decisions, such as products and services.

Business design (page 153): A person with this expertise understands the organisational levers that are required to be designed to give effect to changed experiences and interaction touchpoints. They will be keenly aware of organisational dynamics, culture change and the depth of design required to enact change. This expertise enables organisations to change, which is essential in making impact in complex systems.

Design delivery (page 154): A person with this expertise is a strong program manager and strategic designer. They understand how programs of work are created, and what is required to ensure these are delivered to meet the strategic design outcomes (such as changed experiences). They know how to navigate the executive and decision levels of organisations and broker productive interactions between organisations to achieve results in complex systems.

Action-led evaluation (page 155): A person with this expertise is experienced in specific qualitative and quantitative research and evaluation methods. They understand how adaptive evaluation methods work in the design method and process, and are adept at planning, executing and reporting evaluations that inform iterative improvements to design solutions and define design impact.

TYPICAL COMBINATIONS OF EXPERTISE

There are two ways to think about the design expertise areas:

- as a designer – one person will usually have expertise in several of the core areas

- as a design team – core expertise areas are gathered in different combinations to address different design challenges.

Each of these is considered on the following pages.

As a designer – typical combinations of expertise

Designers who are effective in exploring, innovating and enacting change are usually multi-expert. By way of example, Figure 19 shows the expertise of six actual designers at ThinkPlace. For each indicated area the designer has a minimum of five years' experience. It's not a comprehensive mapping but demonstrates the combinations of expertise that good designers can bring to any design challenge or project. In practice, these designers have the flexibility to operate across many stages of a design project. They also specialise in diverse types of complex system-in-focus expertise, such as compliance and regulation, vulnerable communities and development.

Figure 19: Typical combinations of expertise – designers

	DESIGNER 1	DESIGNER 2	DESIGNER 3	DESIGNER 4	DESIGNER 5	DESIGNER 6
Complex system-in-focus	●	●	●	●	●	●
Co-design leadership	●	●	●	●	●	●
Strategic facilitation	●	●				●
Design research			●	●		
Information and visual design			●			
Experience and interaction design		●				●
Behavioural insights and change					●	
Business design	●			●		●
Design delivery				●		
Action-led evaluation					●	

As a design team – typical combinations of expertise required

By combining areas of expertise in a design team we create a strategic resource for any design challenge. The members of the team constantly learn from each other, testing each other's biases and world views. Figure 20 shows examples of the types of expertise required in design teams for three design challenges. It includes examples of specialised skills, such as agile delivery, that are additional to the core areas of design expertise and may be included in a design team to meet a particular design challenge.

Figure 20: Typical combinations of expertise – design teams

DESIGN CHALLENGE	DESIGN TEAM AREAS OF EXPERTISE	
Design the strategic direction of a complex system	• Co-design leadership • Strategic facilitation • Complex system-in-focus • Design research • Information and visual design	This combination of expertise supports a design of the future that is driven by insight and strategy, with collaborative action across diverse groups to achieve the vision and powerful visual artefacts that share and translate the vision for multiple stakeholders.

Enabling services through digital channels	• Co-design leadership
	• Strategic facilitation
	• Experience and interaction design
	• Complex system-in-focus
	• Design research
	• Information and visual design
	• *Non-core expertise:*
	  – *Technical architect*
	  – *Agile delivery*
		This combination of expertise allows the strategic design of the future service, translated into user-centred terms and interactions. The insights into the way the e-service needs to be structured inform the detailed design, and agile expertise helps translate these insights into digital products that are iteratively delivered.

Organisation transformation to deliver changed experience	• Co-design leadership • Strategic facilitation • Organisational design • Information and visual design • Design research • *Non-core expertise:* – *Change manager*	This combination of expertise ensures that the transformation of the organisation is driven by user-centred insights that frame the change. Strategic design means the organisation has a clear sense of why it is changing and what needs to change – this is expressed through visual blueprints.

A CLOSER LOOK AT THE AREAS OF EXPERTISE

Complex system-in-focus

This area of expertise is about recognising the type of system we are designing within so we can draw from underpinning theories, models and applied theories to understand how we make sense of, intervene in and change systems. It typically includes expertise in areas such as:

- regulation and compliance

- risk and opportunity management

- behaviour-shaping communications

- working with vulnerable communities

- indigenous communities

- public health and other health domains

- education

- the development sector.

In each area – and there are many more than the examples above – system-in-focus expertise brings a literacy that helps us communicate to the various actors in the system. This literacy lends credibility and access to the people in the system who need to take part in the design process, which in turn contributes to our ability to situate and argue the value of taking a human-centred design approach to deal with the issues and challenges they face. This influencing role is instrumental in the reframing of complex issues, and facilitates a pathway for the final design solution to be argued and presented credibly within the complex system.

Co-design leadership

Co-design leadership refers to the expertise of knowing the principles of co-design and how these are applied in a project – how to structure the design project to best meet the requirements of a complex issue, and integrate the four voices of design (page 33). It requires incredible creativity and foresight to anticipate the likely stages and required activities, conduct initial scoping, propose the project and achieve endorsement. This is backed up, as the project unfolds, by ongoing observation and the ability to pivot the process as required to address emerging findings and possibilities. It's not project management expertise, but rather a design-led expertise that embraces the design process as an emergent (non-deterministic) process and brings people along on the journey so they can best co-design and co-create the solutions.

The role inherently seeks to build the capacity of project stakeholders to engage in the co-design approach. Because we take a collaborative approach in design projects, we need to impart co-design skills to senior people, project team members and stakeholders. This may involve targeted training, skilling and coaching.

Importantly, co-design leadership helps alleviate the risk concerns of clients and stakeholders, who may be uncertain that a human-centred approach will work. In the complex issues we work on, we often find risk-averse stakeholders who are reluctant to take an exploratory approach and uncover insights for change. This is the situation where co-design leadership shines, because designers with this expertise are able to talk through the pros and cons and build confidence that the human-centred approach will help solve the challenge. We have often heard clients say they 'trust the process' after they have been engaged by a designer with expertise in co-design leadership.

Strategic facilitation

Strategic facilitation is the expert ability to be neutral, broker fresh perspectives and galvanise action towards a common objective. It draws from foundational facilitation skills that empower, engage, excite, energise and extract best performance from groups of people for design purposes. Further, strategic facilitators are highly expert in strategic thinking and navigation because, unlike other facilitators who are content neutral, they bring expertise in how to drive strategic change and direction throughout the design process.

A designer with expertise in strategic facilitation brings the essence of design to the facilitation experience. This means they are committed to the practice of co-design and injecting the different voices of design. Critically, they give people agency by creating environments that encourage dialogue and build ideas from all participants. They are expert at breaking down power structures by staging activities and experiences that allow participants to experience others' perspectives in a design challenge. A good strategic facilitator will enable all voices and perspectives to be heard throughout a design process.

Expertise in strategic facilitation is highly valued by clients because they often don't have the skills, tools or neutrality to productively engage people with different perspectives, views and goals. This ability to work with opposing views is an essential aspect of designing in complex systems. Gathering these views from the start means the process of understanding the underlying problems, co-creating possible futures and arguing for changes to be undertaken have all involved the key people who will enact the future.

Strategic facilitators are artful in managing information. They are highly visual, conceptual thinkers who can grasp concepts and represent them in workshop settings to extract the valuable wisdom of the group. They use a variety of techniques, including mind-mapping, performance, prototyping and video to capture and reflect the experience of the conversation.

Design research

Design research is a broad area of expertise that draws from a range of fields – such as anthropology, social and behavioural sciences, and psychology – and brings strong skills in research design, planning, execution, analysis and reporting. The critical emphasis in design research is how we locate the user, or the experience of people in the current system, as a core part of what we want to know. The design researcher focuses on the experience of a person to inform the design of the whole.

The premise of design research is that we want to discover, rather than suppose or assume, what end users or clients need or want. In addition, design research helps to empower the end user in the process and invites them to co-discover and co-create the possible solutions. Experts in this area not only apply proven methods but also invent new methods to best discover the human experiences and insights that can inspire design solutions to complex problems.

Design research spans three main goals: discovery, innovation and evaluation.

Discovery means using specific methods to understand what is happening now. We seek to gather a respectful and empathetic understanding of the lives of others. We want to go beyond the standard methods that elicit self-reporting, instead focusing on *in situ* and contextualised settings. Design researchers who are expert in observational studies and ethnographic methods go underneath what is being observed and seek to understand the cultural and behavioural drivers for those observations.

Innovation means using methods that co-create solutions with end users and key people in complex systems, such as intermediaries and people who play an active role in the system. The expertise here is to generate ideas and explore how these ideas might be prototyped. This expertise in prototyping will commonly be paired with other designers, such as interaction designers and organisational designers.

Evaluation means testing and learning about solutions, from low-fidelity prototyping through to more developed design solutions for scaling to make impact in complex systems. The value of the design research methods here is to test early and iterate solutions that meet the needs of the users.

Information and visual design

The expertise to take information and visualise it in ways that evoke insight, draw empathy and inspire action is essential in the design process. Visualising complex data artfully and scientifically can quickly draw attention to the key messages and reveal patterns about the relationships, similarities and differences, changes, and connections between things (such as organisations and processes) that may not be obvious using written or verbal communication alone. Visualising complex data intertwines the goals and priorities of art and science, producing something that is at once aesthetically pleasing, scientifically interesting and factually correct.

Visualisation expertise is applied at all stages of the design process from insight, to innovate, to recommendations for change, and flows into the delivery of new product interfaces and services. It is a highly valued area of expertise because the visual translation of a problem or idea helps to engage the people in the system and provides evidence to act. Further, visualisations can demonstrate the impact and value created from design projects by turning evaluation data into insights of impact, which provides clients with substantive evidence of success.

It's critical that visual designers working in complex system design challenges engage with and understand the content and information they are seeking to represent. Merely rendering a good aesthetic is insufficient to communicate complexity and inspire insights for action.

Experience and interaction design

This is the expertise to translate design research insights into a model, or framework, of desired experiences, which in turn informs the design and implementation of products and services. Designers with this expertise understand:

- how to create a coherent future user experience

- the types of projects that will be required to deliver this experience

- how the future experience will add to the brand and positioning of the organisation (or organisations) delivering the experience

- the goals and measures of success on delivery of the full experience.

This expertise spans multiple roles. The experience design role considers multiple experience layers and interactions, involving frontline staff, intermediaries, trusted agents of clients, and the clients or end users themselves. This is key in a complex system – the ability to appropriately represent a full set of future experiences.

The interaction design role focuses on the way in which users will interact with a specific product or service. The designer translates the human factors and behaviours into product design and features, and works with design researchers in the iterative user testing, design and build of products and services.

Designers with expertise in experience and interaction design are valued by clients because they take the abstract and make it concrete – they turn the vision of a changed experience into something tangible. They engage intensively with users and decision-makers, and can distil what makes a good final design from the opinions and ideas on changes to design features. A great experience and interaction designer will know how to keep focus on the end outcome and result, and manage feedback to avoid over-engineering the final solution.

Behavioural insights and change

To design in complex systems we need to understand patterns of behaviour, and to identify the desired behaviours and behavioural drivers to meet certain system outcomes, such as health outcomes or compliance with a law.

People with expertise in behavioural insights and change draw from behavioural economics, anthropology, sociology and psychology. They take known applied theories and interventions (with behaviour results), fuse these with observed behaviours for a specific challenge, and then co-create, with users, testable interventions to help shift behaviours in complex systems. The ability to understand observed behaviours and what drives people can identify gaps and also, importantly, uncover positive deviant behaviours that we want to emulate across a system.

This expertise contributes to the formation of design research insights and the translation of insights into design decisions, such as products or services. Design decisions may relate to incentives to encourage a group in the community to change their behaviour, the pathways people use to exchange or transfer knowledge, or the willingness of people to adopt a new innovation. When these ideas are prototyped and trialled with user groups, we can use behavioural insights expertise to provide another evidence data set by setting up measurement frameworks to assess the changes in behaviours between control groups and testing groups.

The value of a behavioural scientist in a design project is that they can manage client expectations about the expected rate of behaviour change and related factors that will be evident if a design solution works. They will resist the need for results to be demonstrated quickly, and instead can provide indicators of evidence that solutions are working. This is important because behaviour change is very complex and can only be measured over a period of time – for example, an individual's heightened awareness of an issue may cascade into their contemplation of the issue, which may cascade into a decision to change a personal habit/practice, which may cascade into a fully embraced belief/value, which may then cascade into someone else's heightened awareness, and so on.

Business design

A person with business design expertise understands the organisational levers that are required to be designed to give effect to changed experiences and interaction touchpoints. They design the organisational architectures that align strategy to the new operating business models, structures, processes and capabilities that will enable delivery. This expertise focuses on the pieces of the organisation that need to be redesigned in order to enable desired future experiences. A business designer takes an outside-in, design-led approach. They see that the effectiveness of any business design should be measured in terms of the experience – for clients, key partners and staff of the organisation.

The business designer is keenly aware of organisational dynamics and the culture change required to enact change. They understand how to navigate the decision-making arrangements and equip key decision-makers with clear design artefacts to enable rigorous decisions. The partnership with information and visual designers is critical in this area.

This expertise includes an understanding of cultural norms and how a design project will be activated and supported. The ability to create environments that are open to design research, discovery processes and the co-creation of solutions is key to successful business designers. This is because we enable change-makers to embrace new ways of working that empower staff and stakeholders, and together the new organisation is created. If organisations don't change then nothing changes in the system.

Design delivery

The expertise of overseeing the delivery of programs to achieve desired experiences and products is critical in large organisational settings. A person with this expertise is a strong program manager and strategic designer. They understand how programs of work (priority projects that underpin a desired future experience) are created, and what is required to ensure these are delivered to meet the strategic design outcomes (such as changed experiences).

Designers with this expertise are a powerful partner inside organisations as they can help leaders and implementation teams to focus their effort and ensure the implementation of a well-designed change occurs as intended. They keep the narrative of the expected change alive among staff and key stakeholders. They have at their disposal insights, design artefacts and other designers (such as design researchers, and experience and interaction designers) to communicate the work being done and keep people engaged with the vision of the future.

They know how to navigate the executive and decision levels of organisations and broker productive interactions between organisations to achieve results in complex systems. They are an active agent in the organisation setting, and they enact good governance and reporting to ensure the implementation meets expectations.

Action-led evaluation

The expertise to measure and evaluate is an important part of any design process. This is the way you can ultimately tell if the design was good. In our measurement framework we think of evaluation in four areas:

- **Efficiency** refers to the conversion ratio between inputs and outputs. While it is an important metric, on its own it does not say whether the product or service being developed is fit for the intended purpose.

- **Effectiveness** is about whether the product, service or system is fit for its intended purpose. An effective system is one where the desired future and the actual result are similar. The challenge is therefore to describe the desired future in a way that can be measured later.

- **Experience** is about good staff experience plus customer experience. Staff experience generally will translate into a good experience for customers. A good customer experience will generally translate into the organisation achieving what it is there to do, whether that is a commercial or non-commercial purpose.

- **Ethics** asks whether we delivered in a way that was considered fair, transparent and ethically appropriate by our stakeholders.

The evaluation frameworks that are best suited to complex systems are adaptive and action-oriented approaches. This means the expertise to evaluate design solutions needs to be contemporary and fast-paced. It is live, agile and iterative, feeding back more real-time data points to evaluate the effectiveness of different solutions and interventions. The design evaluator is adept at planning, executing and reporting evaluations that inform iterative improvements to design solutions and define design impact. The design evaluator understands how to co-design the evaluation framework and implementation to suit the nature of the design challenge and proposed solution.

The value of this expertise in a design project for a complex system is that it produces learnings to improve design solutions and insight on how to scale design solutions, which is where impact can actually occur. Strong evaluation helps decision-makers and clients to get necessary funding and resources to scale great designs.

06

TOOLS AND TECHNIQUES

Building blocks for a better future

CHAPTER 6:
TOOLS AND TECHNIQUES

In previous chapters we explored our end-to-end design journey. From the high-level core design diamonds, to the way we approach various design challenges, to the teams of experts we assemble to meet those challenges, we've now reached the most granular level of the design task. The tools and techniques featured in this chapter are the building blocks of our work. They are the way we make our ideas real, transforming them into something we can act upon and modify. They help us break down the problem, dissect it, and co-create something new. These tools may help us to test an idea's potential for value creation, serve as decision-making or thinking aids, or anchor our conversations with the core design team, providing us with a reference point from which to reframe and reorient our thinking. While some tools and techniques are better suited to certain activities, all of them serve to simplify collaboration. We use them because they improve the way we relate to each other. They allow us to cut through dense data and technical thinking and provide us with a common ground to stand on.

Design in complex systems draws on a range of disciplines. In developing our tools and techniques we have modified and shaped practices from these disciplines to serve design purposes, and the descriptions in this chapter should be seen in that context.

We owe a debt to the many specialists in various disciplines who developed the fundamentals of some of the practices we describe. If you wish to know more about these fundamentals, the bibliography at the end of this book lists key sources we have drawn on over the years to develop our tools and techniques.

Some of the practices described in this chapter apply to fields other than design. It's not our intention to present 'text book' or comprehensive descriptions of these practices, but rather to focus on how we have adapted and applied them to design projects. For example, ethnography and social research are specialised fields with widespread applications and are backed by an extensive body of knowledge on their theory and practice. Again, if you wish to know more about these fields, we urge you to read further among the sources listed in the bibliography.

The tools and techniques are arranged into four categories (page 160). We've arranged them in the order that they might be used throughout a project, though we recognise that projects don't always progress in a linear way.

"THE DETAILS ARE NOT THE DETAILS. THEY MAKE THE DESIGN."

CHARLES EAMES,
ARCHITECT AND INDUSTRIAL DESIGNER
1961

In this chapter

Learning deep evokes the maxim 'fail fast, fail cheap, learn deep' and features the tools and techniques we use to investigate the way design can change people's lives. When we use this set of tools and techniques, it's to make design personal: to elevate the user's perspective, to contextualise our understanding, and to fuel connection by building empathy.

- Ecosystem mapping
- Ethnographic research
- Group conversations
- Key stakeholder interviews
- Card sorting
- Affinity mapping

Telling the story refers to the host of visualisation tools and techniques we use to cut through the stark reality of data and inspire people. It's how we make sense of the data we gather, producing a compelling narrative that blends both the tangible and emotional aspects of a user experience.

- Conversation tracking
- Mapping experiences
- Personas
- Business process mapping
- Case studies
- Empathy mapping

Thinking beyond refers to the tools and techniques that help us make a way forward in our work. They compel us to think 'outside the box' and ensure there is a way of actually implementing the change we've co-designed.

- Idea sheets
- Ideation sessions
- Ideas labs
- Scenario planning
- Design camp
- Solution blueprinting

Building and testing showcases the range of tools and techniques that facilitate the process of scrutinising our work. These tools and techniques help us get it wrong to get it right.

- Prototyping
- Usability testing

LEARNING DEEP

Ecosystem mapping

Time needed	1–3 days / 1 workshop
Design System phase	Intent

To understand how change can be made in a particular context, it's necessary to first understand the ecosystem of people and activities that would be implicated in the change. This occurs during the early stages of a project, when we want to know who to consult, who the key people are, and who has an interest in finding a solution to the challenge. Mapping this ecosystem helps us frame the design approach and define a line of enquiry for the design. Moreover, ecosystem mapping reveals the complex interdependencies between and among the various actors (individuals, organisations, cohorts). Because we are complex system designers, we want to identify where our intervention could have the biggest impact. Ecosystem mapping helps us complete the picture and know where to focus our design effort.

There are several common styles of ecosystem mapping, each with its own rules and approaches. It's beyond the scope of this book to explain these but to find out more search for:

- causal loop mapping

- stock and flow mapping

- actor network mapping

- stakeholder mapping.

There are three challenges with ecosystem mapping. First, it's hard to do. Second, even when you go through the effort the map will be wrong because it will always be an approximation with bits missing and inadvertent biases. Third, only you will understand it, others will be bemused.

This is not a reason to give up, but recognise these limitations before you start. Decide how much time and effort you are prepared to put in and map the system to the level of detail your time permits. Ensure you get breadth. It's better to have a wide and shallow map than a map that is deep in parts but omits whole areas of relevance. Finally, don't use the maps as communication devices. Interpret the maps to find the insights that are worth communicating.

For a rapid ecosystem mapping exercise, the kinds of questions we ask are:

- Who are the human agents in the system – individuals, groups, organisations?

  – what value are they seeking?

  – what are the competing interests?

  – what are the common interests?

- What are the non-human agents in the system – for example, features in the environment, legislation, regulations?

- What are the relationships between the human and non-human agents? Or, what things move between people (such as energy or documents).

- What is the supply chain?

- Who has accountabilities and responsibilities in this ecosystem?

- What is the legislative, political and social environment in which these agents operate?

To interpret the map, the types of questions we ask are:

- What are the variables that are of interest? In other words, what are the things that we want to go up or down? For example, we may want to see decreases in the level of poverty, the level of hunger, the level of infant mortality.

- What are the drivers of those variables? Are there specific leverage points?

- How does feedback work in the system? What are the reinforcing loops? What are the attenuating loops?

- What parts of the system are working against each other? What are the critical dilemmas?

To move to action the types of questions we ask are:

- What is the intention for change in the system?

- Who will be affected and how?

- What are the unintended consequences?

- Can we find the simplicity – the other side of complexity?

- Are there a very small number of deep rule changes we can make that will shape the whole system for good?

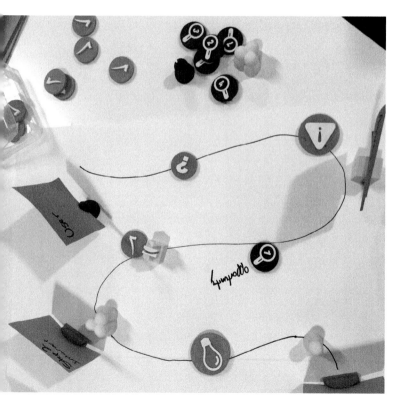

How-to

Typically, ecosystem maps are created in a workshop setting through an intensely collaborative process, though they can be kick-started by the designer conducting contextual or desk-based research before the workshop. Teams are asked to identify the people, processes, policies, and resources and technologies that create the ecosystem. The 'voice of design' facilitates the conversation and maps out the ecosystem according to roles, location (hierarchical, geographic), modes of interaction and so on.

Keep in mind

- Invite people into the mapping who represent the breadth of the ecosystem of interest. A partial map will lead to unintended consequences for another actor.

- People may want to give up part way through because it's hard. Don't give up.

- Conduct ethnographic research (see page 166) so that the map is informed by human experience. Remember that people don't experience whole systems, they experience their pathways through them.

- Never try to stand in the shoes of others. It's not possible. Instead, involve others directly in the mapping. Too often designers in complex systems think that they know better than the people who will be affected by the proposed change. Whether this is prompted by ignorance, arrogance or lack of trust, none are noble attributes of the designer.

- Look for other research. Statistics, studies and reports will ensure that your map leverages what has been done before and adds more.

- All maps and all models are wrong. Some are helpful. There are many different ways to map the system and all will only be a partial representation of the system. Therefore, don't strive for the one accurate map. A number of maps may be more helpful through a variety of perspectives.

"I worked on a project recently where we were asked to define new dementia research priorities in Australia. It was helpful to think about dementia research as an ecosystem which included interdependent stakeholders – not only the researchers themselves, but also the medical practitioners, policymakers, and those who actually experience dementia. Who does someone with dementia engage with? What's their pathway? Who're the stakeholders? Understanding the context in this way was a technique for holding ourselves accountable. Every bit of the project had to feed back to that person, which added a lot of value to the project. If we had only spoken with the researchers, our findings wouldn't have been as honest and true. Throughout the process, we kept in mind that the research is for that person at the centre of the ecosystem, not just today but into the future."

Aimee Reeves
Business Designer, ThinkPlace Australia

Ethnographic research

Time needed	Varies based on content
Design System phase	Explore
You'll also need	📄 Research script
	☁ Consideration of ethical risks
	📝 Participant consent form
	📷 Voice recorder or camera

Design thinking is about grounding design in concrete observations and providing an evidence base to our clients that they can reference when making important decisions. For this reason, human-centred design research has its roots in anthropological, sociological and psychological research methodology. Ethnographic techniques are powerful because they elicit an understanding of how people genuinely engage, value and communicate with each other. A range of tools and techniques can be used when doing ethnographic fieldwork.

How-to

There are many types of ethnography but they all involve being in the context of the people who influence, or are influenced by, a proposed change. This can vary from a multi-month immersion to a few hours of semi-structured interviewing in a person's surroundings as they go about their activities. All of these methods require detailed planning and recordkeeping, and systematic analysis. Ethnographic techniques include the following.

Contextual inquiry: This is a place-based, semi-structured interviewing technique in which the researcher uses the context of the interview as a reference point to provoke data from the participant. The researcher starts with a list of open-ended questions but uses it only as a guide, allowing the participant to respond to the context. This may mean that the participant goes about their normal routine (for example, in the workplace, home, going to appointments) and the researcher asks questions based solely on the actions and surroundings of the participant.

Narrative storytelling: The narrative interview is about achieving a rapport and setting that encourages and stimulates the participant to tell a story about some significant event in their life and social context. This technique allows the participant to tell us their own story in their own language about who they are, why they might have taken a particular course of action, where they have come from, and what sorts of things they want to achieve. We may spend 1 to 1.5 hours with the participant in their home or a location that will help us understand their context, or it may be a shorter part of a discovery interview focused on a particular experience we are interested in.

Shadowing: This is an observation technique in which the researcher joins a person going about their normal routine. This technique works best when the researcher manages the situation to minimise disruption to the natural flow of activity, for example by being a solo researcher and practising rapport and reflexivity with participants. Shadowing can also include participants narrating their activities to elicit meaning, emotion and understanding.

Participant observation: The best way to truly empathise with people is to live the way they do and experience their world first-hand. This form of 'deep hanging out' may be as simple as sitting around in an office or hospital, or it may involve living in another community. This method involves greater levels of immersion with the group of people we are interested in. It requires us to genuinely follow the priorities of another, but to

do so critically. It demands that the researcher build rapport, join in activities and take on a position as an 'insider', while also consciously observing and reflecting. This is a delicate balance because of the level of trust and solidarity involved and the potential for your involvement to significantly shift common behaviours. The researcher must be both the outsider and the insider at all times, wholly participating in the rituals and routines that any other trusted member would.

Video ethnography: Video can be used with any of the above approaches. It's valuable because so much more information can be captured. It can be revisited in the analysis stage and others may see things that the researcher did not. The videoing should be as unobtrusive as possible. Sometimes people can do auto-ethnography using video. A valuable technique is the coupling of video ethnography with 3D playback. This allows presentation of the findings to people who were not there using augmented reality, allowing people to look in all directions at the context.

Digitally enhanced ethnography: Ethnographic research can be extended through the use of digital platforms to collect information from respondents. These platforms can extend the number of people involved, the numbers of regions involved, the length of time people stay engaged and the level of detail that can be gathered. They can also reduce unit cost. Their best application may be with people you have worked with one-on-one to extend the duration of contact. They are not without their downsides, such as recruitment challenges, keeping people engaged and the need to inform people sufficiently to provide meaningful data without skewing results.

AEIOU: Useful to draw out different elements of the whole service experience you are observing. 'AEIOU' (Figure 21) is an abbreviation for:

- Activities – what are people trying to achieve, and what are their main processes and tasks?

- Environments – where do the activities take place and what is the space like? Use all of your senses to take in the setting, and (if you can) write it down before the strange becomes familiar.

- Interactions – between whom and how do interactions occur? These are the building blocks of activities. They can be routine or unique.

- Objects – what tangible elements such as devices or tools are involved? These are building blocks of the environment and are often powerful links to emotion, history and relationships.

- Users – what is their identity? What are the relationships, values and roles of the people involved? Remember that users aren't limited to end users.

Keep in mind

Ethnographic research is different from statistical research. Statistical research allows you to say with a calculated level of confidence that a certain proportion of a population has a set of characteristics. Ethnographic research goes deep and explores meaning, aiming to surface insights, challenges and opportunities. Its rigour is in its systematic application and analysis. Usually, six to ten observations will be sufficient to surface most of the challenges and opportunities. Statistical research goes wide rather than deep.

Figure 21: The AEIOU framework to guide direct observation of an experience

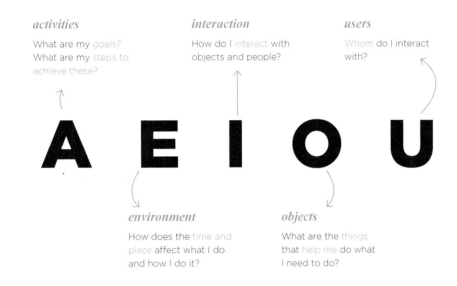

activities

What are my goals?
What are my steps to
achieve these?

interaction

How do I interact with
objects and people?

users

Whom do I interact
with?

environment

How does the time and
place affect what I do
and how I do it?

objects

What are the things
that help me do what
I need to do?

Ethnography is about tracking the candid, mundane and unedited aspects of life. While this may seem like a research technique with very low input from the researcher, on the contrary, ethnographic methods demand focus and strict attention to detail so as not to misinterpret the natural flow of activity.

When taking field notes during observation, try not to make any inferences as to why someone is motivated to do something. This may seem obvious, but it can be surprisingly difficult to document only what is seen and heard, and nothing else.

Contextual inquiry works best when the researcher encourages the participant to articulate previously unarticulated aspects of their experience. The researcher should probe what may seem familiar and overlooked by the participant.

This is sometimes a two-stage process. A stakeholder interview may build trust and rapport. The participant may then agree to being part of an observation or shadowing activity. Using a videographer or taking photos can add depth to observation and shadowing but is only appropriate after a relationship has been established.

"One of the most important tools to have in your belt when you're conducting ethnographic research is to be able to master the pause. Sometimes silence is golden. If you can learn when to pause and allow people to process their thoughts and then articulate them, then you're sure to get richer data."

Darren Menachemson
Partner, ThinkPlace Australia

Leslie Tergas, Design Director and Partner,
ThinkPlace New Zealand

Empathetic listening

Empathetic listening is listening to understand, accessing our curiosity about people while being non-judgemental. This requires deferring your opinion, listening to the whole story, understanding the surface and also the thoughts and feelings beneath the surface.

To achieve empathetic listening we have to slow down, be patient, talk less and listen more, and repeat back what was said to ensure that we don't overlook anything and the person feels heard.

Empathetic listening requires discipline to see through the eyes of the user, and willingness to have beginner's eyes. You must teach yourself to treat every customer interaction as though this is the first time you have ever heard this issue, even if you think you have heard it before.

Group conversations

Time needed	1–2 hours
Design System phase	Explore
You'll also need	📄 Research script
	📝 Participant consent form
	📷 Voice recorder or camera

There is a range of group discussion techniques used to enhance design research. We prefer the term 'conversation' because facilitators are active participants in the conversation, listening, synthesising and provoking. Group conversations vary on several scales: level of facilitation (the degree to which the facilitator participates in the conversation), interactivity (how much the participants engage with each other and actively build on each other's ideas) and flexibility (the degree to which the conversation is anchored in an interview script). For instance, sometimes it's more strategic to facilitate a conversation very loosely, leaving room for emergence and self-discovery that may not occur in a conversation kept strictly to a set of predetermined questions.

These conversations are powerful research techniques because, unlike traditional market research, they take place before a product has been fully designed. Instead of putting a product at the centre of a conversation – with the risk that participants will feel intimidated or constrained by the product – inviting insight earlier in the process encourages a more divergent conversation that brings out insights based on lived experience.

We conduct these group conversations during the 'explore' phase, when we are interested in learning more about what change we could make in the user's life. We prefer to keep group conversations small (at most 4–6 people) to discourage intimidation and promote an environment where every participant feels that they have room to speak freely. Still, group settings tend to favour those who are outspoken or comfortable articulating their honest thoughts to a larger audience. As facilitators, we need to ensure that everyone gets a chance to speak and that they feel comfortable doing so.

How-to

Usually we conduct highly focused 'mini' group conversations with 4–6 members (rather than the more common 10–12 member focus groups). If members are expected to be a random set of users, they can be found through a recruitment agency. Another way to recruit is to use a customer or client listing from one of the partner organisations – this allows you to target more carefully the characteristics of those you seek to understand. The conversation can revolve around a task or set of tasks (as in usability testing) or can simply be a mediated discussion about a particular product or service. The researcher is looking for insights that may not come to light except through group interaction, such as the way members articulate memories or ideas.

- Prepare a set of open-ended questions for the group, but allow these to adapt to the conversation. In this sense, being adaptive means holding the discussion in a setting that allows the participants to have an equal footing (not in a participant's superior's office, for example), using the common language the group uses to describe their experiences, and allowing the group to naturally discuss among themselves (laterally) rather than directly with you.

- Think strategically about the composition of the group. Do you want to have a cross-section of an organisation or team with each key function represented? Do you want to have separate group conversations, one with management and one with general staff? Do you want to separate by certain criteria (such as age, gender, professional rank)?

- Get the logistics right. Find a space with blank walls and room for people to move around. Make sure you have enough butcher's paper, sticky notes and marker pens – avoid ballpoint pens and pencils, as these are difficult for people to read. If it's helpful, provide people with some light pre-reading.

- Begin the conversation how you would a stakeholder interview. The group conversation can be run by a team of two or three people, or by a single facilitator. If the group is run by a team, be sure your roles are clearly defined (lead facilitator, conversation tracker). Spend the first 10–15 minutes building rapport and asking preliminary questions to ensure you and the group are on the same page and a common interest is established.

- Follow the script as loosely as needed to ensure that participants feel free to talk and fully explain their answers. Listen with sensitivity and empathy, probing where the participant is vague, emotional, or refers to an experience without describing it.

- When the group conversation is finished, thank all participants and (if appropriate) distribute an honorarium or gift. When participants leave, there should be an immediate debrief and rough synthesis of insights.

Keep in mind

- Depending on the context in which you are holding a group conversation, be aware of social norms and conventions about group gatherings. In countries where the professional workspace is marked by a strict hierarchical tradition, for instance, the researcher may arrange for superiors to be in a separate group conversation from more junior-level staff to protect the integrity of the data. In other places, dynamics related to gender, ethnicity and more will need to be considered.

- Because group settings emphasise collective knowledge, group conversations often yield perception-based data or data that is riddled with rumours, common myths or shared experiences. This data can be useful, but it's the researcher's job to parse what is useful from what is not.

- Group settings quickly reveal who is more vocal or willing to share their experiences than others, and sometimes this can skew the data. The researcher must be cognisant of the group dynamic to ensure that all voices are heard and the data gathered avoids extremes or exaggerations of any kind.

Key stakeholder interviews

Time needed	30–60 minutes
Design System phase	Explore
You'll also need	Research script
	Consideration of ethical risks
	Participant consent form
	Voice recorder or camera
	Research protocol

Key stakeholder interviews are semi-structured conversations between a designer and the key stakeholders, or influencers, who have been identified (prior to fieldwork) as having a vested interest in solving the design challenge at hand. These interviews gather a range of perspectives on a matter that may yield insights or considerations that would otherwise not have been obvious. The one-on-one nature of stakeholder interviews means that an interviewee may share information that would not have emerged in a different environment. The entire interview can focus on what the interviewee feels, thinks and knows. Giving the interviewee control, time and space to think and respond, and assurance of the value of their perspective, ideally means more honest feedback that can be constructively harnessed. The stakeholder interview is very valuable to sketch the dimensions and perspectives of the ecosystem of interest.

How-to

Identify the research questions that you want to be able to answer by the end of all the interviews, then prepare open-ended questions that could help you elicit answers. Think of the interview as a conversation; don't be wedded to a script or pre-prepared questions. Remain open to visual cues and be aware of topic areas that may be more sensitive or emotive. When conducting a stakeholder interview, be sure to bring the following: a voice recorder, the discussion guide (interview questions), a consent form, and a safety contact (that is, a person who is on-call in case you feel unsafe for any reason). Similarly, it's essential that two people go to any site, especially when it's someone's home, to ensure personal safety. All of this information should be covered in your research protocol.

A research protocol will comprehensively address the following questions:

- Which cohorts will be interviewed?

- How will participant recruitment be managed, and by whom?

- Will payment incentives for participants be paid and, if so, how much?

- How will the information you collect be used?

- How will you record the information you collect (for example writing, audio recording, photos)?

- How will your notes, recordings and similar documentation be stored or destroyed after the research study is complete?

- How will the interviewee's confidentiality/anonymity be maintained?

- Will the interviewee need to sign a legal or privacy consent form?

- What will you tell interviewees in response to questions about the purpose of the study?

- How will you respond if you find out they are engaged in self-harmful or illegal activity?

- How will you respond if you or they feel uncomfortable or unsafe during an interview?

- How will you ensure your safety and limit your risk during an interview?

- Will you interview alone or with a partner or note-taker? Do you have clearly defined roles?

- Who will know where you are at all times, and what will they need to do if you don't get in touch for a few hours?

Rather than reading a list of pre-determined questions, keep the conversation relatively natural by probing where necessary. Stakeholders in a one-on-one interview have an opportunity to proffer differing or more personal views, providing valuable information. By building rapid rapport and creating a safe space for the interviewee to provide their feedback, a designer creates the opportunity to enhance the richness of that information.

There are four key areas to the interview, discussed below. Figure 22 is an example of the first two areas of an interview script.

1. Opening comments – While you don't want to overly script your interview, there are certain elements you will need to make sure you do get word perfect in order to protect the participant and to preserve the client's integrity, such as any legal or ethical aspects. Standard wording can be used and replicated regardless of the topic.

2. Context setting and rapport building – The first 10–15 minutes are for set-up, explanation and preliminary questions. This is an important activity to build rapport and establish a connection with your participant. This is often where they will decide how much they will reveal to you. It's important to be natural and allow the questions to flow as opposed to coming across as 'following a programme'. The participant should feel that you are authentically interested in hearing about their world.

3. Situational storytelling and observation for information gathering – Allow up to two hours for this (prototype testing may take up to 1.5 hours). In formulating this section, you should reflect on the research objectives and associated activities you may be undertaking, and create the framework that will support these activities. You're seeking to deconstruct what people think, do and use, uncovering the underlying factors informing them.

4. Closing and housekeeping – Give the participant the opportunity to ask you some questions. Capture any questions you know you want answered but were not covered earlier. Wrap things up by reiterating legal and ethical aspects, and thank the participant for their time and openness.

Keep in mind

- Be mindful of consultation fatigue. If people have been interviewed a lot before, but haven't seen any progress on what they care about, they may not give you the best responses. If this is the case, consider how else you can gather their views rather than necessarily having to have another conversation with them?

- Stakeholder interviews provide an opportunity to demonstrate a genuine consultative approach that informs an as-yet undecided course of action, bolstered by the advantage of the 'independence' of the researchers, who are not part of the system (they are the 'voice of design' – the independent brokers).

- Look out for presumed constraints. Ask why the stakeholder considers them constraints and how they have interacted with them.

- Consider the need for recording and transcribing.

- Test carefully anything that is unclear to ensure you have properly understood what someone has said.

- Schedule one to two hours between stakeholder interviews to allow time to document and reflect on the interview, before the nuances are crowded out by the next event.

"In my view, the secondary value of stakeholder interviews is in credibility building. If you are attempting stakeholder interviews as part of a design project, you are presumably either attempting to understand a challenge, gathering ideas and insights that may inform a solution, or creating a change. Regardless, you will only be successful if you have credibility and support from stakeholders."

Steph Mellor
Senior Executive Designer, ThinkPlace Australia

"Interviewing is both a science and an art – it's all about balancing those two. It can be difficult to master the art of conversation while also being in control of the way you're investigating and interrogating someone's perspective (lived truth). The informant is the expert, and the art of conversation is making them feel that level of security. It's about maintaining a balance between letting the conversation flow naturally, while also recognising where to probe and tease out concepts. You have to be both inside the conversation and outside of it at all times."

Carlyn James
Design Researcher, ThinkPlace Australia

179

Figure 22: Research script example

THIS IS THE FIRST PORTION OF AN INTERVIEW SCRIPT WE USED FOR A DIGITAL PRODUCT DESIGN CHALLENGE – TESTING AN EMAIL/DIGITAL COMMUNICATIONS CHANNEL FOR THE COMPLETION OF OFFICIAL TAX DOCUMENTS.

INTRODUCTION (5 MINS)

Hello, my name is _____ and I'm from a research and innovation company called ThinkPlace. We are working with the Department of [agency name] to better understand the way users interact with their processing system.

Now before we begin, I'd like to take you through this participation agreement. It has important information you need to know and agree to before we begin.

Hand them a copy of the participation agreement, and then read through it, expanding as necessary.

Now, take a few minutes to read through it if you would like, and ask me any questions if there's anything you don't understand or are unsure about. Once you're done, if you could sign it, I'll sign it too.

If external and due for a participation payment, and asks how or when they will be paid, say: The research company will contact you to arrange your payment in the next day or two. It usually gets deposited directly into your bank account.

CONTEXTUAL INQUIRY / BUILDING RAPPORT (5–10 MINS)

INTERVIEW QUESTION (ASK THIS)	PROBES
Tell us about yourself? How long have you lived/worked in this area?	• General Information • Understand their usage of internet/computers in the home/at work
How complicated would you say your tax affairs are? Tell us about your tax lodgement last year? What interactions have you had with the tax website?	• Understand the user's perception of the complexity of their tax affairs • What sort of things do you claim? What sort of investments do you have? • Have your circumstances changed since last year? • Describe experience (good & bad points)
What is your first impression when you see this email? Without opening the email, what would you expect to see once you open it?	• Initial perception of the email • Whether they trust the sender • How likely would they be to open it / treat it seriously
(Once they have opened the email) What are your first impressions? What stands out immediately on the page? What would you do next? Would you click on these buttons? What would you do if you had questions or needed help?	• First impression (whether it meets their initial expectations) • What is the main call to action • How likely are you to click on the button? Would you have any concerns about clicking on it? • What would you expect to see? • Would you reply to this email?
(Ask specifically about each section box) What do you think this means? Where do you think it would take you if you clicked on it? What do you think of the style of the email? What does it remind you of?	• Comprehension of each of the boxes • Visual style • Brand impression

Card sorting

Time needed	30 minutes
Design System phase	Explore
You'll also need	Pre-printed cards
	Barcode scanner (optional)

Card sorting is a user experience design tool that informs the information architecture (that is, the navigational structures, images and language) of a website or any group of items or concepts. It is a form of quantitative mental modelling or mind mapping that collects data across many users. Card sorting seeks to identify the way people classify items: what is common and what is different. We all categorise differently so there is usually no completely right or logical answer to a card sort.

Each item is typically represented by an index card and participants are asked to arrange the cards into groups that make sense to them. Participants then name the groups in their own words, offering insights into the language they associate with various concepts. Consider organising food in a supermarket. Does sweet chilli sauce belong in the 'sauces' aisle or in 'Asian foods'? Does a protein snack bar belong in the confectionery/candy aisle or the health foods aisle? People will sort these things based on differing opinions and associations.

Card sorting offers a quantitative way to engage people's input into structuring information. It helps to find the best compromise between different mental models and approaches where a structure or grouping may not be obvious or strictly logical. Structuring information according to a client view of what is logical or technically correct, or that uses legislative or technical terms, may actually make it difficult for users of that information to find what they need. This is a common problem with websites, organisation structures, procedural manuals, product inventories and many other large information structures. Card sorting mitigates this problem because it is inherently user-centred and based on the premise that organising and accessing information is subjective. This is a tool we often use in the 'explore' phase to inform our design of a service or information architecture.

How-to

Conducting a card sort has several basic steps. Start by defining a list of topics or items that you want the participant to categorise. The list needs to be relatively long (around 50 items), but not so long that the participant is overwhelmed. These topics or items are written onto index cards (or similar) or entered into online card-sorting software. The cards are shuffled and given to the user, who then arranges them into categories. For this activity, avoid time pressure but ensure that the participant does not spend too much time. Typically, the first thought is the most useful as it represents natural thought patterns. The categories should be labelled by the user in their own language. If the cards are physical cards, the researcher should take photos and document the categories for statistical analysis later. If the card sort was conducted using sorting software the statistical analysis can occur immediately. There are two standard kinds of card sorting:

- Open card sorting – The users are given a set of cards with various topics or items on them and are asked to put them in groups. They are then asked to create labels for the groups. In open card sorting, participants have freedom to create and name their own groups with no restrictions. Open card sorting is usually used at the start of an information architecture exercise, to identify broad patterns that can be used to start creating a draft information structure.

- Closed card sorting – The researcher provides the categories and the users try to sort the cards within those predefined groups. The participants' sorts are then compared to see

whether the draft structure is accurately representative or if people find the draft structure to be unintuitive. Closed card sorts are usually conducted to test a proposed or draft structure. However, other methods such as tree testing (see below) usually give better results because they are task focused rather than sort focused.

Tree tests are similar to closed card sorts but entail giving users predefined groups. Rather than being asked to sort items, they are given a broad set of tasks. This sorting exercise demonstrates whether the user can match their tasks to the structure given to them.

Keep in mind

- Work with the client to refine the information you will put on the cards to avoid any topics or items for which users may not have a reference point.

- A card sort should help you define which structural questions to ask. For example, you may want to test whether 'buying a laptop' belongs in a finance category or an IT category. In your card sort you would include some finance-related cards, some IT-related cards and the 'buying a laptop' card, and watch where the participants sort the 'buying a laptop' card.

- Card sorting is useful when a set of unknowns has already been identified – that is, if you have the information you need but don't know how to organise it. If you don't know what information you need, a card sort is not ideal. You can use other methods to identify the information you need, and then proceed to a card sort activity.

THREE COMMON MYTHS ABOUT CARD SORTING

Myth 1: Coming up with the information for the cards is easy to do. This first step, getting the correct information, is key. Find a way to assess whether you have the right information, as this will encourage richer, more accurate data. For example, you may need to ask the client for a diverse list of items (including locations, departments, services, user cohorts and others) that would relate to various sites.

Myth 2: Card sorts produce only subjective, user-specific information. Card sorts reveal patterns in how different people group and order information that counterbalances the subjective feedback that users provide. The quantitative data produced in a card sort activity, when graphically represented via cluster mapping or a dendrogram, is a powerful and relatively cost-effective, simple way of gathering large amounts of user experience data.

Myth 3: Card sorts are paper-based, making them less efficient than other tools. Card sorting can be done on a computer (using card sorting software) or hardcopy cards (printed onto perforated paper cards). Using a barcode scanner, the cards can be logged into the software and graphically represented. Additionally, depending on the setting and the user's capability (such as their comfort with screen technology), the researcher may determine that a paper-based card sort is optimal.

Affinity mapping

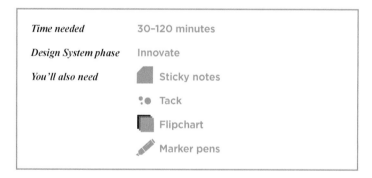

Time needed	30–120 minutes
Design System phase	Innovate
You'll also need	Sticky notes
	Tack
	Flipchart
	Marker pens

Clustering large amounts of data is an important part of analysing any field data and is a core skill for designers in complex systems, because of the need to determine where to place boundaries, identify similar lines of thought and decide which thoughts and ideas should be 'parked' and returned to later.

Affinity mapping illustrates the range of a problem, reveals similar experiences among users, and identifies areas for further study. It also helps designers synthesise their findings and organise their thoughts before writing insights reports or completing data visualisations. The labels that are developed through affinity mapping reveal major themes, which then inform the ideation process.

To analyse the data collected from the 'explore' phase of a project, we use affinity maps or diagrams to cluster information. Affinity maps are interactive workshop tools that group data based on natural relationships and aid understanding of a significant amount of diverse information. They help organise patterns and themes, allowing us to move from a literal interpretation of the data to a higher level of interpretation. This technique can help make sense of the information and bring new aspects into sharp focus.

How-to

First, cluster the data into loose groups of similar topics, concepts or ideas. Then label those groups and revisit the map to see if, with the addition of the labels, some of the topics, concepts or ideas belong in different groups.

1. Ask participants to write down research insights, observation notes, or any generated notes (such as from brainstorming), using a separate sticky note for each.

2. Have participants introduce each note to the group.

3. Begin to cluster related issues together as they are presented. Sort all the notes into the different broad categories.

4. Look for key observations and name them by category. Ask the group to propose changes to the clusters as you go.

5. Review the groupings aloud, looking for patterns or themes.

6. Once the notes have all been presented and categorised, give the patterns or themes meaningful names.

Keep in mind

- Affinity maps are an interactive design method, so they can be incredibly useful in keeping workshop participants engaged and allowing them to come to their own conclusions.

- Affinity maps may appear very complex and disorderly when they are in a paper-and-pen form; they should be captured (via photo or other documentation) and then documented for easy consumption.

- There's never a correct way to group information and there's never a complete way of grouping information during an affinity mapping exercise.

TELLING THE STORY

Conversation tracking

Time needed	1 workshop and 1–2 days after
Design System phase	Any phase

Conversation tracking is a technique to visually capture a conversation from a workshop or meeting and present it back to the participants in a highly consumable format. The artefacts (the conversation trackers) are typically a PowerPoint document that captures the conversation in text and visual form. As well as providing a snapshot of the thinking on the day, conversation trackers should provide an early synthesis and sorting of the complexity of the problem. They are a simple technique, yet are hugely valuable in communicating not just the content of the day but the context in which the conversations occurred.

The conversation tracker serves as an interim document and information source for the development of future artefacts. A conversation tracker helps those who were in the workshop to 'relive' the conversations they had, and for those not in the workshop it helps them to understand the thinking and rationale for decisions made. Conversation trackers capture the complexity of a conversation in a way that provides clarity, using information and visual design. The final decisions and agreed designs are documented in more formal documents, such as blueprints, strategy papers and discussion papers. Capturing the detailed thinking as it occurs means that producing the final product is faster, true to the contribution of participants, and provides a rich source of material from which to synthesise the final design.

How-to

Documentation is a very important part of any workshop because it ensures the final designs accurately reflect participant input. While it's possible to go from the conversation to a finished product, this is not recommended; though you can use the conversation tracker to develop a paper or strategy document at a higher level of abstraction. There are several levels to consider when thinking about conveying a conversation.

- Level 1 is the conversation itself. The only way to experience this is to be in the conversation.

- Level 2 is a transcript. This includes detailed minutes of everything said, or a video or recording of the event.

- Level 3 is a conversation tracker. The conversation tracker captures the flow of conversation but applies an initial synthesis by identifying clusters in the conversation and summarising the issues and themes that are discussed. The conversation tracker will also take emerging diagrams and concepts and sharpen them. It will include images of the whiteboards so participants can refer back to what was said. The audience for a conversation tracker is often limited to those who attended the workshop or will use the output.

- Level 4 is a standalone strategy document or design document. This type of product is often built from multiple inputs including desk research, quantitative analysis, workshops and interviews. It is a fully synthesised product; it has a logical structure and would make sense to anyone. Blueprints, strategic plans, detailed designs, program plans and position papers are examples of such outputs.

Conversations tend to start messy and become clearer as they progress. This should be reflected in the way the conversation is captured on the whiteboard – starting with mind maps, working towards models to create understanding, and ending with defined lists of tasks and time schedules. In mind maps, the facilitator is chunking key pieces of information that are linked.

This allows the person creating the conversation tracker to create these groups of information in adjoining bubbles or boxes. When capturing, key verbatim quotes can be used to capture and convey powerful experiences and opinions of participants. Full sentences that stand alone and make sense for any reader should be used, and always include a few sentences about a page or an image to give it narrative. Be sure to include photos of the whiteboard, the work, and the scene in the document. Photos enable the reader to understand the context of conversations and connect with participants.

Follow these steps to build a conversation tracker.

1. **Set up a template in advance** with blank pages as per the agenda. Setting up the blank pages will give you an idea of how the conversation may flow and save time in writing up the conversation later.

2. **Document the conversation as it progresses.** During breaks in the conversation, work with the facilitator to communicate where the product is going and make sure it aligns with what the project needs. This is also an opportunity for quality assurance of the capture of the conversation. Aim to work in real time with the conversation. Listen actively – ask yourself: What is the key theme here? What are the points

and counter points? How can I summarise this point? How does this point link to what has come before? Where is this conversation going? Use this line of self-questioning to lift your own comprehension skills and concurrently stay in the conversation and document it. By the end of the workshop, raw content should have been captured for each section of the conversation tracker, ready for refinement afterwards.

3. **Work closely with the facilitator.** To capture the conversation at this level of synthesis, the person who is tracking and the facilitator who is whiteboarding work together interactively and in parallel. Both people should understand the purpose of the workshop, the final design and product that is being developed, and how the conversation will proceed (the agenda). The facilitator may capture a few concise words on a whiteboard for speed and flow of the conversation, and the conversation tracker should include sentences and prose, as well as visual design. The facilitator clusters the key parts of the conversation on the whiteboard and the conversation tracker uses this as a cue for how to capture the conversation in the tracker. A good facilitator will be able to visualise the content of the conversation in different ways, such as through models and diagrams that explain a complex concept easily.

4. **Send the conversation tracker to workshop participants within 48 hours.** It's best to finish the conversation tracker the day of the conversation or the next day, as this keeps the conversation fresh in your head and the quick turnaround allows participants to review the material while the conversation is still fresh for them. It also provides a sense of pace, energy and direction for the participants, and helps the project to move ahead more rapidly. As mentioned above, have a good first draft of the conversation tracker completed at the end of the workshop. Use the time afterwards to improve language, add more detail and proofread the product.

Keep in mind

- This technique is not something that's meant to be laboured. The turnaround for this deliverable should be very quick, within 48 hours of the conversation taking place.

- If you are conversation tracking, your relationship with the facilitator is important. Having a good rapport and being responsive to each other's speed in terms of tracking and facilitating makes the task much easier.

- If it's a workshop, don't forget to take candid photos to include in the final product – these can't be recreated, so don't miss the opportunity.

- The document should be highly readable and a compelling and coherent read from beginning to end.

- Use whole sentences and only include dot points that are the continuation of a sentence.

- Ensure you understand each of the clusters and have these recorded in a coherent format.

- Avoid excessive words. A good conversation tracker is concise and efficient in language.

"Intellectually and artistically capturing conversations and packaging them this way has real value. It keeps the conversation live and present even after it's happened. It can also serve as a reminder or a new call to action for participants."

Kerstin Oberprieler
Executive Design Manager, ThinkPlace Australia

Figure 23: Conversation tracker from ideation workshop

INSIGHTS

THINKPLACE

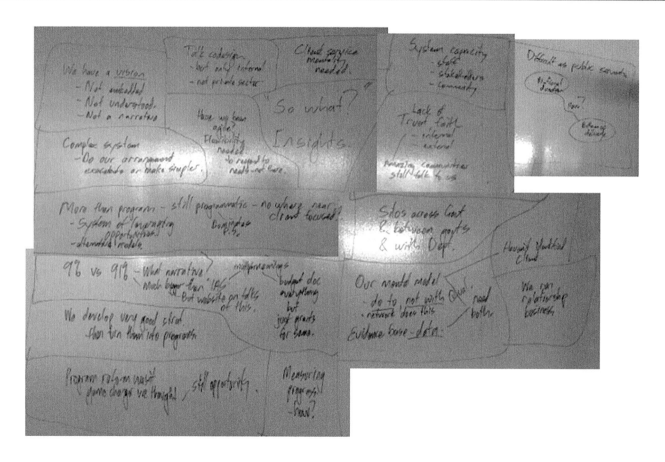

INSIGHTS

THINKPLACE

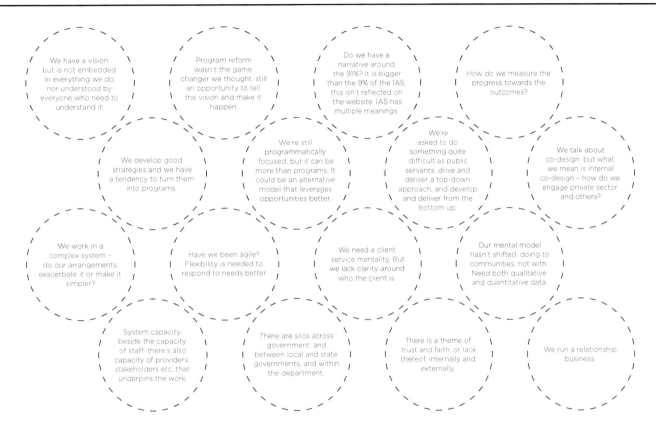

We have a vision, but is not embedded in everything we do, nor understood by everyone who need to understand it.

Program reform wasn't the game changer we thought, still an opportunity to tell the vision and make it happen.

Do we have a narrative around the 91%? It is bigger than the 9% of the IAS, this isn't reflected on the website. IAS has multiple meanings.

How do we measure the progress towards the outcomes?

We develop good strategies and we have a tendency to turn them into programs.

We're still programmatically focused, but it can be more than programs. It could be an alternative model that leverages opportunities better.

We're asked to do something quite difficult as public servants, drive and deliver a top-down approach, and develop and deliver from the bottom up.

We talk about co-design, but what we mean is internal co-design – how do we engage private sector and others?

We work in a complex system – do our arrangements exacerbate it or make it simpler?

Have we been agile? Flexibility is needed to respond to needs better.

We need a client service mentality. But we lack clarity around who the client is.

Our mental model hasn't shifted: doing to communities, not with. Need both qualitative and quantitative data.

System capacity: beside the capacity of staff, there's also capacity of providers, stakeholders etc. that underpins the work.

There are silos across government, and between local and state governments, and within the department.

There is a theme of trust and faith, or lack thereof, internally and externally.

We run a relationship business.

Figure 24: Conversation tracker from ideation workshop

DESIGN PRINCIPLES

THINKPLACE

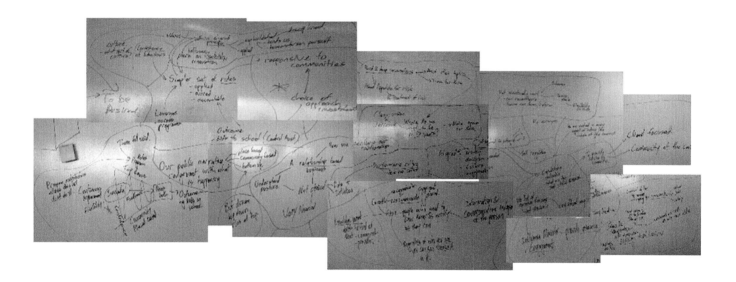

DESIGN PRINCIPLES

If you went and bought a shrink-wrapped version of the Architecture, what are the top features you would look for?

☐ We are a **relationship-based business**. The Architecture understands how we configure our relationships and our posture in a nuanced manner, and not be clinical.

☐ We work with a mix of **data and intuition,** the Architecture articulates how this aligns with an outcome-oriented culture.

☐ The Architecture fosters a **clear vision and narrative,** creating space and discipline for us to address the question: where do we want to be in ten years?

☐ The Architecture makes time for **hard and deep** conversations about a range of topics, particularly our appetite for and treatment of risk.

☐ The Architecture should **not be structurally inert;** we can reconfigure when contemporary issues arise, teams can form and reform, with shared outcomes; a stable base structure with flexibility within this.

☐ The Architecture is aligned to achieving the **outcomes and priorities** of the Government, and allows clear and pithy communication and assurance to the Minister.

☐ The Architecture is **self-regulating,** If you fall outside of the Architecture, you are brought back into it.

☐ The Architecture aligns the **capabilities** of staff with the outcomes that need to be achieved. It refocuses how we think of ourselves, beyond programs and teams etc.

☐ No more acronyms.

☐ We are involved in every aspect of our clients' lives. Our Architecture is **client focused** with the community at the centre.

☐ Ultimately, the Architecture facilitates the successful **delivery of services** to the community.

☐ We need to shape the conversation with stakeholders, ensuring that we value-add, constantly **engage and change the conversation.** Co-design both internally and externally, and leverage the 91%. Know when to get in, and when to get out.

☐ The Architecture ensures people who need to know things, know those things as quickly as possible. Information and **communication** needs to be independent of any one person. People need to understand where they fit.

☐ The Architecture facilitates good **knowledge management,** where everyone can tap into information, particularly about place: community and providers.

☐ The Architecture fosters the ability and confidence to do **both top-down and bottom-up** policy making that is responsive to communities. It ensures authority and credibility.

☐ The Architecture ensures our **public narrative** and statements are congruents to reality, we operate in the way we say we do.

☐ The Architecture works within an articulated set of **shared principles** based on our humanitarian pursuit; that we understand, adhere to and apply in our work. This should provide the stable base for flexibility and innovation.

☐ **The Architecture fosters a principle-driven culture, with consistent behaviours, a simpler set of rules that we can leverage across programs, and is applied, owned and accountable.**

Mapping experiences

This section looks at three types of experience maps. Each has a different purpose.

- **Life event maps** are useful to organise a mapping exercise but not a good way to describe a user experience because life events are usually years (if not decades) apart. A person experiences starting school, finishing school, getting a job, buying a house and so on over many years. While these events can be helpful to identify the different experiences to map, a map of life events is not in itself meaningful from a user experience perspective.

- **Journey maps** are developed for different identified cohorts, such as students starting university, people earning between $5 and $20 per day living in a rural area, or people seeking employment. The maps can be developed using either ethnographic techniques or workshops with relevant participants (the ethnography-inspired maps will be of higher quality). Journey maps show the whole user experience within the scope of the map. For example, if the map is about finding a job, it will show engagement with educational institutions, recruitment agencies, potential employers, friends, mentors and others. The purpose of a journey map is to make big step improvements and changes to a whole experience.

- **Service touchpoint maps** show the experience, or interaction, with a specific service or product. Their purpose is to improve a product or service by revealing the triggers, pain points, emotional responses and more that the user has in relation to a single experience.

Journey maps and service touchpoint maps can describe the current or future state; the latter is built by modifying the former.

Figure 25: Mapping experiences

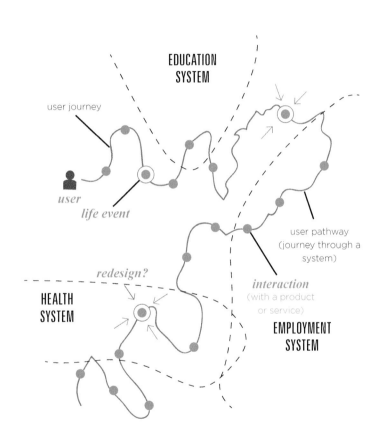

EDUCATION
SYSTEM

user journey

user

life event

HEALTH
SYSTEM

redesign?

interaction
(with a product
or service)

EMPLOYMENT
SYSTEM

user pathway
(journey through a
system)

Life event mapping

Time needed	1 day / 1 workshop or the output of a multi-week piece of ethnography
Design System phase	Explore

Over the course of a person's life, they will have many experiences that shape who they are, how they see the world, and the situation that they find themselves in. When we document this set of events we create a life event map. Visualising the user experience is a crucial component of the co-design process. Life event maps can achieve this by producing visually compelling representations of real people's experience.

Journey mapping

Time needed	1 day / 1 workshop or the output of a multi-week piece of ethnography
Design System phase	Explore

Journey maps show a typical user's typical sequence of daily activities, relevant to the experience that is being studied. For example, if we are designing a service for pregnant women in rural areas to remotely interact with a physician, we may want to map out events such as:

- attending reproductive health classes at school

- getting married

- becoming pregnant for the first time

- becoming pregnant for the second time

- having a pregnancy complication.

None of these events are part of the service being designed, but understanding that they happen (and what happens when they do) is critical context for the designer when thinking about how the service integrates with the user's context and life experience. The impact of journey maps is that they allow us to construct a story that reveals the range of positive and negative elements of a user's experience. Visualising these elements helps us locate particular elements of a user's service experience where we could make a positive change. Essentially, journey maps help us accomplish the following:

• building user insight

• guiding project teams

• informing overall design integration.

Building user insight

Some of the questions that journey maps help to answer are:

• What are the larger goals of the user?

• What are they actually trying to achieve?

• What are their pain points?

• Where can we intervene in the system to make improvements for society or the environment?

• How can we add value to the services we offer and make them work better for our users?

• What drives our users to make the choices they do?

These questions can't be answered using just transactional data. Journey maps combine several methods of ethnographic research and other qualitative methods, coupled with quantitative data such as usage statistics, to build a more comprehensive picture of the user experience.

Guiding project teams

Project teams can draw on this picture to aid decision-making about the services that their project will impact. Some of the questions project teams can ask are:

• How can we offer the product or service that users actually need?

• How can we make it easier for users to achieve their goals?

• How will this decision impact on user's goal?

Journey maps provide a baseline of what users expect from a service and what currently frustrates them about the level of service they receive. By interrogating the journey maps, project teams can make better-informed decisions about their project, which will ultimately positively impact the service that users experience.

Informing overall design integration

Individual project teams can contribute to the overall user experience but they are not in a position to know the overall impact on the user. The user's experience is influenced by many other factors in the ecosystem that surrounds them, and by different departments combining to rationalise interactions with individuals. Therefore, to achieve significant changes requires a different level of abstraction from the maps. This is inherently more difficult and less predictable. The timeframes of the changes when working at this level will be multi-year, possibly five to ten years. The number of stakeholders will significantly increase and could include multiple governments, organisations, groups and individuals.

How-to

The process of developing a user pathway is iterative. It starts during the 'intent' and 'explore' phases and continues through the exploration of concepts.

1. Be clear about the journey you are mapping. A clear definition of the scope should be formulated from the 'intent' definition.

2. Ensure the user is the source of your knowledge. The user is an essential part of the process; inviting their knowledge gives authenticity to the final product. There's a number of ways you can involve the user. For the most value, you need to go out and observe and interview the user in their environment. This provides you with first-hand data about the experiences you want to understand and map.

3. Collaboratively develop the journey map or pathway through workshops. Run a co-design workshop with representatives from across the organisation and users themselves to draw out the different aspects of the research findings. This is essentially prototyping the experience. You may need to run several workshops to work through different types of user and different scenarios, and to innovate new types of maps.

4. Document the journey maps. The initial result of the workshops can often appear messy and may seem illogical. This is why a session is needed to make sense of the information and package it in a visual, presentable, logical story. Experience is different across different services and there is no single way of representing it. Therefore, you

need to take an iterative approach to finding the best way to visually represent the pathway so that it is meaningful to the intent and outcome for the work. Journey maps will generally include significant events that impact the user and how the user experiences those events (for example, feelings of frustration or satisfaction).

5. Test the journey maps with the users. You need to know: Have we got it right? Have we captured the most common experience in the map or pathway? Have we emphasised exceptions or critical failure points in the experience?

6. Communicate the journey map or pathway. In most scenarios of the co-design methodology this will be done through the program blueprint. Journey maps and pathways are powerful and should be utilised as an important communication, engagement and alignment tool.

Bringing journey maps to life

When developing a journey map (Figure 26), focus on key aspects of a person's experience:

- what the person does

- what the person uses

- what the person thinks

- what the person feels.

By understanding and displaying all these elements together we get a rich understanding of where the person experiences points of satisfaction and pain, and the way in which they respond. This allows different viewers to see experiences with fresh eyes and identify areas of opportunity where they can create impactful change, while continuing to do things that are working well.

Note that journey maps can describe the current state or the future state. The future state map is built by modifying the current state map to a desired future experience based on findings during the research phase.

Keep in mind

- Journey maps can incorporate a wide range of materials and resources – photographs and quotes can be used to enhance the voice of the user within the journey map.

- Having a wide range of experience in your team allows for richer layers of perspectives when viewing user experience.

Figure 26: Early sketch and finished journey map

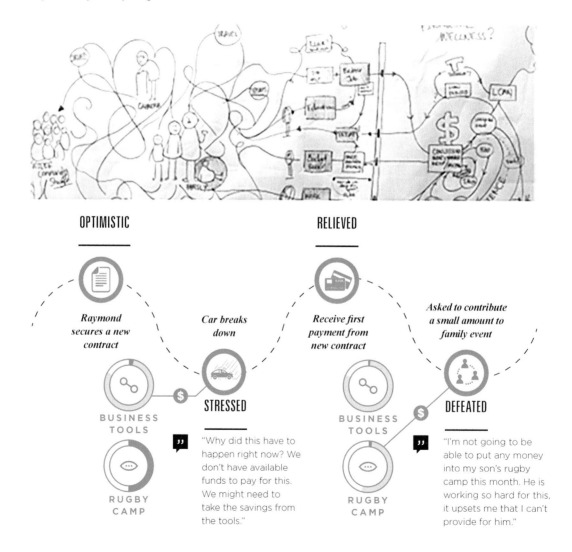

OPTIMISTIC

RELIEVED

*Asked to contribute
a small amount to
family event*

*Raymond
secures a new
contract*

*Car breaks
down*

*Receive first
payment from
new contract*

**BUSINESS
TOOLS**

STRESSED

**BUSINESS
TOOLS**

DEFEATED

**RUGBY
CAMP**

"Why did this have to
happen right now? We
don't have available
funds to pay for this.
We might need to
take the savings from
the tools."

**RUGBY
CAMP**

"I'm not going to be
able to put any money
into my son's rugby
camp this month. He is
working so hard for this,
it upsets me that I can't
provide for him."

Service touchpoint mapping

Time needed	1 day / 1 workshop and user testing sessions
Design System phase	Explore

If journey maps show what happens in a person's life outside the scope of the service, then service touchpoint maps focus on what happens when the user is interacting with the service. The premise of pathways was first expressed by Professor Richard Buchanan (2008), who understood that no one person can experience a whole system such as the revenue system or the health system – it is too complex. Rather, a person experiences a journey through the system and their experiences are defined by the particular pathway they take and the interactions that occur. Like journey maps, service touchpoint maps are tools developed in response to in-depth research findings resulting from the 'explore' phase. Service touchpoint maps detail each interaction that users have with the service.

For example, if we consider once again the service for pregnant women in rural areas, we can trace interactions along their service touchpoints such as:

- registering for the service

- having their first remote interaction with a doctor

- meeting with a doctor to set up a mobile phone-based care plan

- receiving an SMS reminder about an upcoming vaccination.

All of these interactions happen within the bounds of the service, but it's important not to confuse a service with a product. While part of the service may be (say) a mobile phone application, other parts may include filling out paper forms, having face-to-face interactions with a doctor or even the woman discussing something with her husband.

A project team has many factors to consider when designing: what is the legal and policy imperative? what is possible from a technology perspective? how does this fit with existing technologies? how does this fit with other agencies or third parties? what is viable for the organisation financially, structurally

and for staff? how does this initiative interface with work others are doing?

With all of these considerations it can be easy to forget that the activity is ultimately directed at users – users with rights and responsibilities. The user pathways and user insights are intended to guide project teams to improve the user experience, reduce inefficiencies and increase the achievement of program outcomes.

How-to

When developing service touchpoint maps, refer directly to your design research. The map should be a composite of real user experiences, demonstrating what the user was trying to achieve within the system and how their journey played out. The foundation of every pathway should be the users themselves, so be sure you have a grip on who the user is and their motivations.

1. **Start by considering the user's original goal.** This goal might change throughout the process, but focus on the user's original intent.

2. **Define the trigger.** Why is the user on this journey? For example, the trigger may be emotional (dissatisfaction with

a product) or logistical (perhaps there's a sense of urgency because a user is leaving the country and wants to get their passport renewed).

3. **Consider the journey itself – what are the touchpoints?** Where are the sites of interaction between the user and the system? These could range from face-to-face contacts and call centre contacts to accessing a website.

4. **Identify the pain points.** These will be the touchpoints in a user's pathway where there is a negative transition in their emotional state.

5. **Enhance.** Wherever the touchpoint mapping ends, the designer should ask provocative questions about how to enhance the user's pathway. The map should help us to locate exactly where we can have the biggest impact in changing a user experience for the better.

6. **Add abstraction.** Once the pathway itself has been defined, add several layers of abstraction with quotes from the user, emotional reactions to each touchpoint (using emoticons or colours, for example), and photos or icons to add depth.

Keep in mind

- Initial pathways can be sketched out using sticky notes or a walkthrough (role play or scenario-driven) process. The first level is always the process – these are the actual steps the user took. The second level of abstraction tracks the user's emotional response to their experience. At each step in the process, what was their reaction, how were they feeling? The next level of abstraction might be a set of quotations from the user reflecting their immediate thoughts about the journey they are having.

- A pathway may be created to display the experience of an individual user or to generalise across a cohort of users (a 'macro pathway').

Personas

Time needed	1–3 days / 1 workshop based on ethnographic research
Design System phase	Innovate

Personas represent archetypal users of a product or service, embodying their needs, goals and activities. A persona is presented in the form of a narrative that conveys the needs of an archetype user within a particular context and scenario.

Personas offer a meaningful reference point for understanding who you are designing for. They help you tell the story of what life is like for a user today and what they would wish it to be like in the future. This allows you to explore the 'what ifs' in between. Personas also bring research to life for designers and team members who didn't talk directly to users, helping the team keep the users in sight during solution development by developing empathy for the various types of users and the way they make decisions, the pain points they face, and their priorities.

Personas ultimately lead to design objectives that inform the 'Innovate' phase of the Design System methodology. We create personas based on themes that emerge through our ethnographic research, which ensures that we deliver positive experiences based on accurate depictions of our end users. The benefits of using research to inform this tool are twofold. First, it ensures our design direction and decisions are influenced by real data taken from real users, rather than incorrect or inaccurate assumptions about users' individual experiences. Second, by placing our designers in the field to conduct research, we create a deeper level of empathy and understanding, which allows us to create a richer user experience overall. Personas can be presented in a range of ways, from simple documents through to active characters. In East and West Africa, we have developed personas using animation and role play, and in other contexts workshop participants have 'played the role' of the personas, inviting the core design teams to ask questions, engage with them, and observe their experience. These techniques enable livelier discussions and deeper engagement with the lived experiences of real people.

How-to

The key elements of successfully creating a persona are:

- data collection and analysis – this should result from in-depth user research with priority user segments

- creating engaging persona descriptions – personas represent a group of users, so when developing the details of your persona, be sure to capture as many cross-cutting insights from your research as possible

- presenting meaningful scenarios – be sure the scenarios are realistic and reflect actual experiences documented in your research

- acceptance of the persona by the design team and stakeholders – invite stakeholders to put themselves 'in the shoes' of the users and accept their experiences.

Personas are created only after conducting in-depth user research. When analysing the data, look for themes and patterns that can be drawn out – what characteristics are relevant to specific users? what characteristics are universal across user groups? From this analysis you can form rough groupings of personas, naming or classifying them according to their most prominent characteristic ('the helper', 'the risk taker', 'the newbie'). Once the rough groups are determined, you can begin developing details about the personas you are going to create.

- Personas should be developed collaboratively.

- Personas are based on evidence. Gather relevant user information you have uncovered through research, including information from existing research/studies and information the team has uncovered from user research.

- Determine what information is relevant – this should be guided by your intent – to help you focus on fleshing out information that is relevant to the specific work.

- Sketch out the persona based on what you know:

  - who they are – give them a name and a description

  - what they do – give them a profession and describe the activities they do

  - based on your design parameters flesh out any information that is relevant (for example, if the project is about an online service, include information on the persona's technology and interaction preferences)

207

- include demographic information if it exists (this is optional)

- add a picture to give a face to the persona name.

• Consider role-playing an interview with the persona to make sure you have captured the right kind of information.

• Organise persona information in an easy-to-read, logical format. You want to create the highest visual impact, using clever visual design to show the reader which elements of the persona to focus on. The reader should actually feel as though they have a complete picture of the individual and can use the persona as a reference point during the ideation process.

Keep in mind

• Is your persona based on real research or assumptions? Personas are evidence-based tools and should reflect users' lived experiences. There can be a tendency for people to cook up personas in their office. Personas not based on research have no value.

• Is there enough rich information to build empathy between users and stakeholders?

• Will other designers and stakeholders be able to understand the narrative you are presenting?

• As a way of enhancing the client's perspective, personas can be visually arranged along a spectrum to allow the stakeholders to see how users compare along various scales (for example, from low risk-aversion to high risk-aversion).

• Be careful not to fall into the trap of creating shorthand stereotypes instead of personas.

• Personas are built through the blending of common experiences and are useful in wrestling with the complexity and breadth of user experience, but it should be remembered that through the process of finding themes and commonalities, some important experiences can be missed. Designers should be conscious of not losing sight of the real people behind each persona.

Figure 27: User pathway for persona (rural healthcare worker in Ghana)

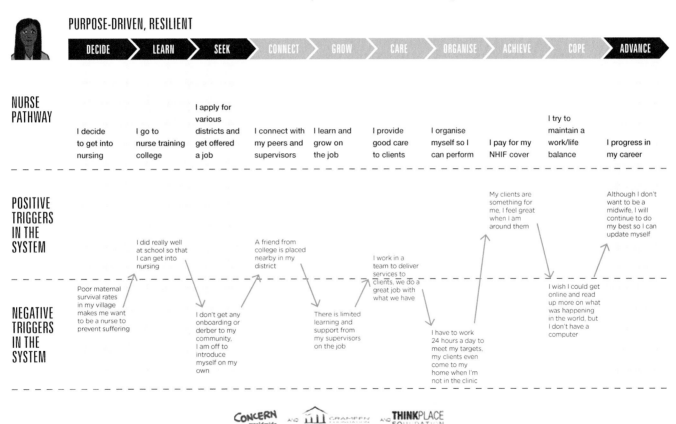

NAANA'S PATHWAY THROUGH THE SYSTEM

This is a typical user journey through the health system as a purpose-driven and resilient Community Health Nurse – Naana's persona

PURPOSE-DRIVEN, RESILIENT

DECIDE → LEARN → SEEK → CONNECT → GROW → CARE → ORGANISE → ACHIEVE → COPE → ADVANCE

NURSE PATHWAY

I decide to get into nursing | I go to nurse training college | I apply for various districts and get offered a job | I connect with my peers and supervisors | I learn and grow on the job | I provide good care to clients | I organise myself so I can perform | I pay for my NHIF cover | I try to maintain a work/life balance | I progress in my career

POSITIVE TRIGGERS IN THE SYSTEM

I did really well at school so that I can get into nursing

A friend from college is placed nearby in my district

I work in a team to deliver services to clients, we do a great job with what we have

My clients are something for me, I feel great when I am around them

Although I don't want to be a midwife, I will continue to do my best so I can update myself

NEGATIVE TRIGGERS IN THE SYSTEM

Poor maternal survival rates in my village makes me want to be a nurse to prevent suffering

I don't get any onboarding or derber to my community, I am off to introduce myself on my own

There is limited learning and support from my supervisors on the job

I have to work 24 hours a day to meet my targets, my clients even come to my home when I'm not in the clinic

I wish I could get online and read up more on what was happening in the world, but I don't have a computer

CONCERN worldwide ᴬᴺᴰ 🏛 GRAMEEN ᴬᴺᴰ THINKPLACE FOUNDATION

Business process mapping

Time needed	1 day / 1 workshop / 1 week
Design System phase	Any phase

A business process map is a tool that maps the key actors, activities, handovers and pain points within an organisation's business processes. It can be used for a variety of ends, such as achieving a consistent practice and experience in the future, streamlining processes that don't work efficiently or highlighting antiquated practices that no longer make sense.

A business process map can help to crystallise a clear picture of the overall effectiveness and efficiency of an organisation's operational processes. Sometimes a map simply helps people understand how many steps there are in a process, often revealing that seemingly simple processes have multiple, sometimes convoluted steps. More often it goes deeper, giving those who participate in the process better visibility and a stronger understanding of the process – for example, how many people and systems are involved, the number of handover points, how effective it is, how efficient it is. It provides clients with the evidence to make decisions.

Business process mapping is different to experience mapping: the former provides a cross-cutting look at the business process (and the specific activities within that process), while the latter illustrates the human experience of the process. A business process is deterministic, that is, it is repeatable and predictable. It is within the scope of an organisation. But a user experience is not deterministic – there are so many other factors involved it can't be predicted.

There are a few key components to represent on a good process map:

- Roles – Who does something in the process? What do they do? What do they need to support them? What is the outcome they achieve by doing their activities?

- Triggers – What triggers the process?

- Activities – What activities do the roles undertake?

- Decisions – What decisions change the flow of the process, or lead to an alternative process?

- Process flows – What is the workflow?

- Information flows – What information flows are involved?

- Systems/tools – What systems and tools are used to aid workflow or communicate information?

- Legislation/policy – What pieces of legislation give rise to powers that are exercised in the process? Where are delegations exercised?

How-to

1. To develop an effective business process map, start by running a 'BPM 101' session with your core design team and relevant stakeholders.

    – Introduce the concept of roles. What does the role do? What do they need to support them? What outcome do they achieve?

    – Introduce a legend for using sticky notes. Assign different colours to activities, decisions, systems/tools/forms and legislation/policy.

2. A handy activity for getting people in the right mind space is to run a brief simple process, such as 'order a coffee' or 'post a parcel'.

    – Farm out the simple processes to small groups. Ask each group to think of the roles involved; think of the activities they undertake; think of the decisions that might change the process flow; and think of the systems that are used (such as a point of sale system).

    – Give the groups five minutes to talk through the process and make a simple process map.

    – Get the groups to present back and critique the process. This is a good time to talk about scope, the differences between decisions and considerations, and give helpful feedback that will help the participants understand how to build a bigger process map.

3. Now focus on the process that is at the centre of the workshop. Develop the **intent** of the process.

4. Get the group to generate a list of the **roles** involved (sometimes this list will need to be rationalised a little).

    – Ask the group to nominate themselves for wearing the 'hat' of a certain role and write the role statement for that role. What does the role do? What do they need to support them? What outcome do they achieve?

- Have the participants run through their role statements and ask the group to provide feedback and fill any gaps.

5. Ask participants to write down all the **activities** that their nominated role performs. Write one activity per sticky note, using a verb and a noun.

6. Start laying out the map horizontally on a table (a table is better for the group dynamic than a wall).

- Ask the group to lay out their activities.

7. Actively work through the map, from the trigger to the conclusion, sequencing the flow of activities and the handovers throughout the process.

- Do a pass for activities. Check that decisions have been included and that the flows from these decisions make sense.

- Do a pass to look at the systems and tools used.

- Finally do a pass to understand where legislation, regulation and policy comes in.

Once you have done this you have a map. But that is only part of the puzzle. The next thing you need to do is use the map:

- Are you streamlining the process? Then look for inefficiencies and talk about how you can increase efficiency.

- Are you trying to establish consistent processes undertaken at different locations? Tease out the differences in how each region undertakes the process and agree on the best practice.

- Are you trying to cut duplication? Look for activities that are duplicated and decide on the best role to undertake that activity.

Don't be afraid to ask 'why' – why does the process work this way? It's not good enough to say 'because it has always been this way'. Think of how the process should best work today.

Keep in mind

- Be clear about why you're doing it – it may be to cut duplication, pinpoint inefficiencies, increase effectiveness, or another end.

- Make sure you have the right people in the room – those who know all of the handovers and other operational aspects of the process, as well as those that know the policy.

- Business process maps can be used in concert with experience maps to provide a more detailed, human-centred portrait of the user experience.

- Business process mapping can be a sensitive issue for some actors who are used to their way of working. It's important to engage those actors early and regularly to ensure their concerns are considered.

- If the business process map reflects the current state, ensure it captures what people actually do rather than what they are *supposed* to do. If the map is meant to reflect a future state, it should show what people are expected to do when change is implemented.

- Allow time for validation of the maps. Often there are areas that require clarification outside of the workshop environment.

"We can show an executive the sheer amount of activities their staff actually do. The executive may look at the map and say, 'Oh, I thought that person did 5 things – it turns out they do 25 things!' It might show how longwinded the user experience is and how much double handling there is."

Mark Thompson
Senior Executive Designer, ThinkPlace Australia

Case studies

Time needed	1 day
Design System phase	Explore

Case studies are in-depth profiles of a single user, group or situation over a period of time. Because they are based on evidence gained through ethnographic research, case studies add richness and depth to clients' understanding of the user. They can be used at the front-end of a project as an empathy tool.

Case studies can be used in conjunction with other tools, such as personas, service touchpoint maps and journey maps, that paint a more vivid picture of the user experience. Case studies give clients the evidence to make certain decisions – they are a cross-cutting way of putting the reader in the user's shoes.

When constructing a case study, think of it as a vignette rather than a long-form resource on the user condition. Keep the case study specific, tying back only to the data provided by the user – there should be no analysis in a case study. Using punchy facts, quotes and statistics provides the reader with quick reference points.

How-to

1. Decide on the level of empathy required from your reader before you commit to writing. Ask yourself, 'How in touch with the users' reality is the reader?' This will guide decision-making on how many case studies should be developed. The more a person reads, the more drawn into the users' world they will become.

2. Listen back to the entire interview recording before committing pen to paper. Take notes on the person's story and bring it together in a logical order, perhaps chronologically.

3. Write an unedited response to what you have heard. At this stage, recording your impressions as completely and naturally as possible is more important than organised structure.

4. Read your case study and delete any emotive words. It's not the job of the case study writer to give opinions or lead the reader to an emotional conclusion.

5. Edit your work. In an ideal world you will be working with a colleague who can do this for you. The distance between a writer's ego and an editor's pen is very valuable. If you are the only person working on the project, try to remove all bias

from your work. Make your language as concise as possible, delete any repetition and ideally keep the case study to one page to keep reader attention.

Keep in mind

- Case studies can be very time-consuming to produce and are best produced by a team – a writer and an editor.

- Emotive language will sabotage your work as readers are rightly mistrustful of overly emotional writing and interpret its use as manipulation. Instead, maintain objectivity – present facts without seeking to create a particular response.

- Keep it short and focus on the salient points to avoid losing valuable substance.

- Case studies have incredible value in situations where a client wants an empathy tool that lends itself to depth, rather than breadth, of understanding. If you're trying to convey breadth, a persona or other archetypal representation of a user cohort is preferable.

Empathy mapping

Time needed	15–30 minutes
Design System phase	Explore
You'll also need	Sticky notes
	Tack
	Flipchart
	Marker pens

Empathy mapping is an interactive workshop tool that succinctly captures the user experience across several sensory layers. This tool can be used during an interview to help process what the user is saying, or after the period of data collection to help find patterns across user experiences. Empathy mapping gives participants the opportunity to understand, visually, where their research insights are weighted. Empathy maps can be helpful before user research when segments have been identified, but they are most useful after research has been conducted and data has been analysed and synthesised.

The empathy map is typically arranged into four quadrants: do, use, think, and feel (Figure 28).

How-to

1. Print out a large-scale empathy map (using the four quadrants) and hang it in your workshop studio.

2. Assign a colour to each user the team will be interviewing or to each user persona.

3. Encourage workshop participants to write research insights on individual sticky notes and place them in the appropriate quadrant. This is a high-energy, interactive activity.

4. Participants either present each insight or rapidly place them on the map and let the facilitator present the insights in each quadrant separately.

5. Look for meaningful patterns and clusters of information. Draw out specific themes that will inform the development of design criteria and concept sets.

Figure 28: Empathy map

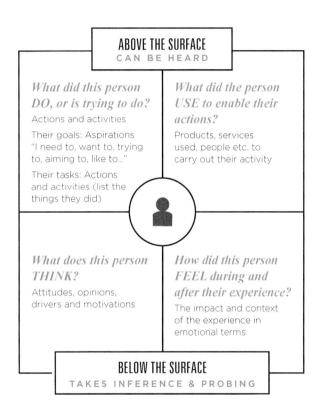

ABOVE THE SURFACE
CAN BE HEARD

What did this person DO, or is trying to do?

Actions and activities

Their goals: Aspirations "I need to, want to, trying to, aiming to, like to..."

Their tasks: Actions and activities (list the things they did)

What did the person USE to enable their actions?

Products, services used, people etc. to carry out their activity

What does this person THINK?

Attitudes, opinions, drivers and motivations

How did this person FEEL during and after their experience?

The impact and context of the experience in emotional terms

BELOW THE SURFACE
TAKES INFERENCE & PROBING

Keep in mind

- This tool is very effective at kick-starting the ideation process, so encourage participants to go for quantity and use the exercise to document minimally edited thoughts. Ideas don't have to be complete or polished.

- Hang the map in a high-traffic area of the workshop studio. When participants pass by in the course of the day they can stop and digest the ideas in their own time. You might leave a marker and a stack of sticky notes near the map to encourage participants to actively build upon others' ideas throughout the day.

THINKING
BEYOND

Idea sheets

Time needed	30 minutes
Design System phase	Innovate/Explore
You'll also need	Tack
	Marker pens

We use idea sheets to progress a conversation where ideas are generated but not properly documented. These sheets allow participants to transition thoughts and ideas (which may be hindered by individual uncertainties, hesitations or ambiguities) into something that can be shared, socialised and enhanced. Idea sheets contain three generic questions (Figure 29) that are meant to guide participants' thoughts and capture ideas. Ideas differ from concepts in that they may be unrealistic or dependent on other ideas to be viable.

Using idea sheets encourages participants to download ideas in a minimally edited way. No ideas should be discounted at this stage, since someone's emerging idea may spark someone else's stroke of genius. This tool lends itself to gamification (such as introducing a time constraint or awarding a prize for the highest volume of sheets completed). Idea sheets generally prompt the use of a decision-making technique

(such as dot voting or clustering) that leads to concept shortlisting and development.

How-to

1. Print about four to five times as many idea sheets as you have participants.

2. Within a set amount of time, have the participants record as many ideas as they can. The time constraint encourages participants to go for quantity.

3. After time is up, the facilitator asks participants to cluster and vote for ideas (to provoke further thinking on those same ideas), and discuss how to develop those ideas into concepts. Often this means shedding some ideas that are deemed not desirable or feasible and combining others that would be more effective as a single concept. Expect to go through multiple iterations to generate viable concepts.

Keep in mind

- Print plenty of idea sheets so participants' thoughts are not interrupted by having to ask for more. This tool intentionally favours volume over content.

Figure 29: Idea sheet template

IDEA SHEET

One idea per sheet, please write legibly!

1 **Name it**
Give it a memorable name (5 words maximum)

2 **Describe it**
Briefly explain the idea

3 **Defend it**
Describe why it is important

Ideation sessions

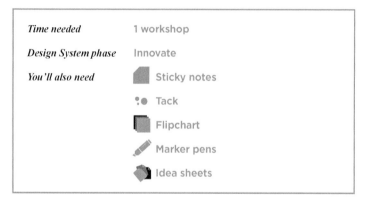

Time needed	1 workshop
Design System phase	Innovate
You'll also need	Sticky notes
	Tack
	Flipchart
	Marker pens
	Idea sheets

An ideation session is the generation of new ideas driven by insights informed by research into the user experience. We run ideation sessions to support the core design team's efforts to generate transformative, actionable solutions rapidly. When running an ideation session, it's critical to allow for separate phases of divergence and convergence. The process of moving from divergent thinking to convergent thinking maintains the momentum of the session.

Ideation sessions are closely linked to several other tools and techniques: generating design criteria, rapid prototyping, idea sheets (tools used within ideation sessions) and evaluating

solutions. Supported by rapid brainstorming and other interactive activities, ideation sessions push people further into exploring their own ideas. These sessions are rigorous and demanding, helping people not only actively build on ideas but also develop them into concepts, evaluate them against certain criteria, and imagine how those components would be embedded in practice. Ideation sessions are meant to provoke, inspire and stimulate people's creative thinking.

How-to

Use the following steps to plan and run an ideation session that caters to the needs of the client and the project at hand:

1. agree the purpose or goal

2. establish criteria for ideas

3. generate many ideas

4. provide feedback and disrupt to improve the quality of ideas

5. select ideas with potential

6. deep dive to develop ideas

7. seek user experience early

8. evaluate ideas against the ideation criteria and the extent to which they strike a balance of what is desirable, viable and possible.

Tools and techniques you may need when conducting an ideation session include:

- idea sheets (Figure 29) to rapidly generate ideas

- feedback at pace (use sticky notes)

- voting for ideas that have the most support from the group

- deep dive sheets to provoke questions to build ideas

- rapid prototyping techniques to transform ideas from concepts into real, tangible solutions that can be evaluated.

Keep in mind

A successful ideation session creates inspiring concepts that meet the design criteria, are desirable, viable and possible, and have the potential for transformational change, with a strong sense of ownership.

Ideas labs

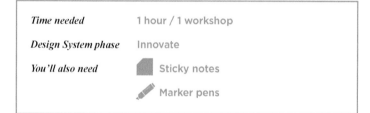

Time needed	1 hour / 1 workshop
Design System phase	Innovate
You'll also need	Sticky notes
	Marker pens

Ideas labs arrange the workshop space into 'laboratories', evoking a sense of exploration and discovery. They encourage participants to think divergently and yet still in a structured way.

The workshop space should be arranged in a very specific way to encourage the greatest collaboration among participants. Each ideas lab should encourage participants to think in a different kind of way. Our standard ideas lab framework has three ideas labs:

- challenges / orthodoxies
- harnessing trends
- leveraging existing ideas.

How-to

1. Prior to the workshop or while the participants are on a break, arrange the workshop space into several 'labs'. Use partitions, whiteboards or tables to encourage participants to gather around each ideas lab (Figure 30).

2. Assign each ideas lab a separate theme (or 'discovery lens') and provide a description sheet for each lab setting out its theme, questions and process (Figure 30).

3. Give participants 30 minutes to explore and ideate in each lab. The ideation session will have as many rounds as there are ideas labs.

4. Ask participants to move to the next ideas lab, refine and discuss ideas that have been generated and explore more ideas.

5. At the end of the session, ask participants to identify the most radical ideas and use these to build concepts.

Keep in mind

The facilitator should stay engaged, even though this activity is relatively 'hands-off'. Continue driving the conversation and pushing participants to explore, develop and enhance their own ideas.

Figure 30: Idea lab set up and description sheets

Room set up and process

THINKPLACE

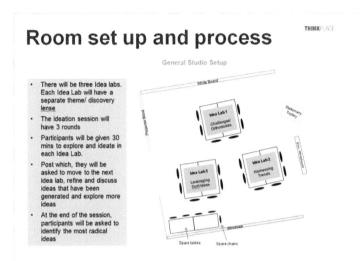

General Studio Setup

- There will be three Idea labs. Each Idea Lab will have a separate theme/ discovery lense
- The ideation session will have 3 rounds
- Participants will be given 30 mins to explore and ideate in each Idea Lab.
- Post which, they will be asked to move to the next idea lab, refine and discuss ideas that have been generated and explore more ideas
- At the end of the session, participants will be asked to identify the most radical ideas

1 — IDEAS LAB

CHALLENGES / ORTHODOXIES

- Reflect on some of the challenges / orthodoxies or conventional beliefs in relation to your organisation (10 mins)
- How can we turn those challenges into opportunities? Imagine alternative ways of doing things (10 mins)
- Reflect back (5 mins)

2 — IDEAS LAB

HARNESSING TRENDS

- Reflect on some of the major changes / trends in the external environment, e.g. technology, demographics, lifestyle, regulation, etc... (10 mins)
- Reflect on the major trends in your industry / space – here, internationally
- What are the implications and how can we leverage on that? (10 mins)
- Reflect back (5 mins)

3 — IDEAS LAB

LEVERAGING EXISTING IDEAS

- Reflect on current ideas that your organisation is currently exploring or doing (10 mins)
- How can we leverage what is currently being done? (10 mins)
- Reflect back (5 mins)

IDEATING: WARNING SIGNS

Risks

- Ideas can feel marginal unless they can be synthesised and fed back properly – being able to articulate the desired shifts, feedback and disruption from the group is necessary to push ideas beyond the obvious.

- Not starting early enough – seek feedback from the user early to provide the team with genuine criticism.

- Not thinking systemically – develop frames and tools that force systemic thinking.

- Big change can be scary – make ideas concrete by prototyping and explore the implications of change early with user testing.

Challenges

- Managing risk – use a 'parking lot' concept to park surfacing risks that can be addressed later.

- Ownership – generating transformative ideas collaboratively increases the ownership and understanding of both the problem and the emerging solution.

Scenario planning

Time needed	Varies (1 workshop – several months)
Design System phase	Any phase

Scenario planning is a powerful technique used in strategy design to imagine possible futures and evaluate potential strategies. It's an important activity when designing interventions in complex systems. Leaders standing at the edge of today need to make decisions that ensure they and their organisations are well positioned for the future.

Scenario planning is not about attempting to forecast one future – it's about imagining several possible and plausible futures. For the scenarios to be useful they should involve factors that could have a big impact on the client's organisation but which the organisation has little control over – for example, how community attitudes to privacy might change, the relative strength of one national economy over another or the impact of a new technology.

Scenario planning can enhance an organisation's capacity to:

- perceive, interpret and respond to change
- influence others
- test strategy
- learn and prepare for the future.

Figure 31: Planning for possible futures

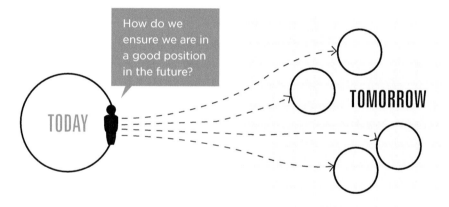

The scenario planning approach is attributed to the oil company, Royal Dutch Shell. In the 1970s it was generally believed that oil prices would only ever continue to rise. This popular view considered the increasing scarcity of oil but not other factors such as international relations and alternative sources.

The strategists at Royal Dutch Shell took a different approach. They asked, 'what if the upward trend in the oil price stopped or reversed?' Shell did not predict the price would reverse but ensured that its strategy would be robust if such a scenario occurred. This meant they were less leveraged into oil assets and better prepared to respond in the event of an oil price reversal. Being ready for this scenario strengthened Shell's position in the global business.

Since then many organisations have used this technique to mentally prepare for possible futures and ensure that their strategy would be robust in a range of futures. For example, a large regulatory agency in the Australian Government wanted to understand the systemic drivers of change for that agency over the course of the next ten years. One of the people who worked on the project recounts:

> We interviewed regulation thought leaders around Australia to understand the future landscape. Our findings about the context presented very interesting possible future dynamics for the agency and its industry to consider, as a community. Using the research, we helped the agency develop four scenarios based on the critical uncertainties – globally connected versus globally disconnected and a regulatory profession that was out of touch versus cutting edge. The agency presented the scenarios and then led a conversation where they could ask 'where do we want to be? How do you want to develop into the future as a professional community? And how can we help you?'

Scenario planning adds discipline and structure to what is a multi-faceted and complex conversation. It allows you to structure thinking about the factors that may drive change. By understanding these factors in the form of provocative narratives about the future, you can obtain a glimpse of what may eventuate and help decision-makers develop their strategies for the future.

How-to

1. Understand the context

Creating scenarios requires expertise in the context in which you are working. Gather people with diverse expertise in the different levels of context, and have a discussion with them about the trending factors in each context. Different layers of context could include:

- global level – what are the political, economic, social, technological and environmental changes that will impact the organisation's future?

- sector level – what are the trends in the sector that will impact the organisation's future?

- local level – what are key factors driving change within the organisation and its partners that will impact your future?

2. Identify the critical uncertainties

After gaining a deeper understanding of the context, factors that present critical uncertainties will emerge. An important question is which trends or instances, should they occur, would have a dramatic impact on the organisation or what it is seeking to do.

List the top uncertainties. The key is to identify the two most critical uncertainties.

Critical uncertainties are:

- things we can't control

- variables that are equally plausible

- variables that can go up or down

- factors that will have a big impact on the future context

- factors or variables that are not related to each other – in other words, the movement of one has no effect on the other. For example, if you were a solar panel provider two variables that may affect your business are the price of raw materials and level of community concern about climate change. These are largely unrelated variables (although everything is ultimately related). By contrast, if your organisation provides employment services, then level of unemployment and level of difficulty in finding a job are too closely related to give useful scenarios. If we strategically plan with dependent variables, then we are actually only mapping one conceivable scenario that's viable, which results in scenario planning that is not helpful to inform strategic planning.

Identifying the two most critical uncertainties is possibly the most important step of the process as it frames the scenarios that will be created. The two critical uncertainties act as the axes that set up the four quadrants for the scenarios (Figure 32). Name the axes and clearly show that they are on a spectrum – for example, low vs high.

3. Develop scenario narratives for each quadrant

Each quadrant (Figure 32) framed by the axes signifies a scenario. Use the axes to develop a provocative scenario describing:

- How did this scenario arise? Using clear language, tell a compelling story of how the organisation developed from today's state to the future state. Present a seamless transition between now and the future to describe how this possible alternative future arose directly from today's reality.

- What is it like in this future?

Give each scenario an interesting name to make them memorable and easier to refer to.

Be clear how many years into the future the scenario is, based on how far into the future the strategic planning is looking.

Developing the scenarios is a great group activity. Having different groups develop each quadrant and then present their own scenarios is a highly engaging process and creates a powerful shared understanding of what futures may eventuate.

4. Develop strategy

The key question is: 'What strategies might we employ to steer us towards our desired future state and away from our least desired state?' Collect all the responses for each scenario and bring them together. You might find that there is some commonality in the strategies, which will help to inform your strategic direction. Organise the individual responses into strategic themes or areas of strategic focus. These are then further broken down into initiatives, goals and key performance indicators. Given how the future could play out, the strategy can be used to navigate what can be done today to ensure the organisation is in a strong position in the future.

5. Test strategy in the scenarios

Now that the four scenarios have been developed, and the strategy is also developed, it's possible to test the strategy in the different scenarios. Scenario planning uses a term borrowed

from aerodynamics research: wind tunnelling. This is where the power of scenario planning really becomes apparent. Scenarios act as a conceptual wind tunnel to stress-test the strategy in different environments.

We brainstorm what is necessary to 'survive' in each scenario. This forces us to think about how the organisation might fare in alternative future environments and exposes areas of weakness and strength. It allows leaders to evaluate assumptions and decisions and course-correct.

As each scenario has the same starting point (today's current state), it's likely that similar responses will emerge, despite different future endpoints. These responses will be used to develop the organisation's strategy.

Figure 32: Scenario planning – quadrants and levels of uncertainty

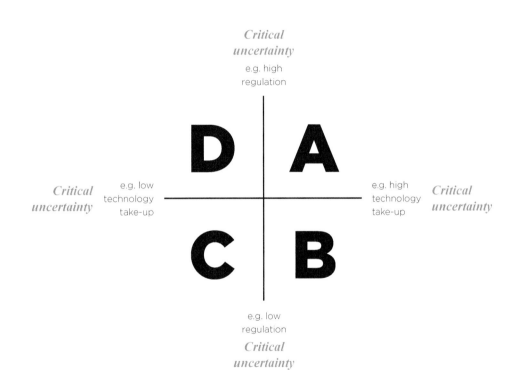

Keep in mind

- It may be tempting to create multiple permutations of critical uncertainties and develop a whole series of scenarios to consider. However, keep in mind the power of having only a few. Too many scenarios will bring back complexity, and defeat the purpose of helping the decision-making process.

- Because a scenario is essentially a story about how trends and factors play out in the future, there's an important element of storytelling. Scenarios should be narratives, making them easily understandable by any relevant stakeholder.

- Scenario planning is a form of 'what-if' thinking, so when brainstorming, capture a wide range of possibilities.

- Scenario planning is one part of the strategic planning process. It needs to be strongly informed by quality understanding of the context. It's important to get the critical uncertainties right, as they underpin the scenario planning process, which in turn strongly informs the strategic direction the organisation will take.

- Don't fall into the trap of thinking you have actually predicted the future – scenario planning is not a crystal ball!

Design camp

Time needed	Full day to multi-day event
Design System phase	Any phase

Design camps are intensive workshops designed to rapidly create a vision, plan or design for a large, complex change. They are a foundational technique unique to ThinkPlace and a cornerstone of how we co-design solutions to complex problems. They are intended for any group engaged in strategic planning processes; early stage design of major policies, programs, or initiatives; major capability or operational reforms; and tackling macro-level and intractable issues.

A design camp:

- brings together a diverse group of people – usually from a variety of levels, areas and across organisational boundaries

- defines the problem or opportunity from a range of critical perspectives

- rapidly and iteratively generates a vision and a clear understanding of the shifts required to get to the desired future state

- deep dives into the activities or initiatives that need to be done to achieve the vision

- considers the integrated plans through a range of lenses, such as risk, timeframe, change management, constraints, stakeholder management and others that could make the difference between success and failure

- delivers a clear, agreed way forward, and a plan to take on the issues and deliver the outcomes.

The questions that a design camp may seek to answer include:

- How will we administer a major new policy regime?

- What is our investment agenda for the next three years?

- What should a new capability look like, and what is the plan for establishing and embedding it?

- What is the plan and high-level design of a change program or reform program?

- How should a new citizen-centred service work, and how should it integrate into the service landscape?

- How should enterprise case management work in an organisation?

- How can we transform a complex social issue?

How-to

Every design camp is unique, but structurally follows the same five-step process.

1. **Form:** First, form a small, core design team specifically for the purpose of organising the design camp. This team typically comprises key external and internal stakeholders.

2. **Define:** Work with the team to define the design camp intent and design the process to achieve it. Ensure you are addressing the right problem; if there is ambiguity around the problem then that should be the first part of the design camp.

3. **Prepare:** Prepare for the design camp, customising all aspects to reflect the specific needs of the organisation. This will often mean preparing templates with questions in advance that will guide the design camp. These may be prepared using large format paper or posters. It may also mean gathering materials for a prototyping session. A key part of the preparation for a design camp is identifying a facilitator and a person to create conversation trackers (page 188) to work together during preparation, during the design camp itself and after it. A big design camp will need a number of support resources.

4. **Gather:** Participants gather to attend the design camp and spend 1–3 intense days designing and planning the future. Participants will be provoked and challenged, leaving with a sense of accomplishment and confidence in understanding how to achieve their future state. To give a sense of difference, don't hold it at a standard work location and don't run a standard 9 am to 5 pm day. It could be longer or shorter (or split), but ensure every detail is designed.

5. **Create:** Create a compelling record (conversation tracker) reflecting the outcomes and the discussions that led there.

Keep in mind

The 'campsite' where the design camp happens is just as important as what happens. Although you could hold a design camp anywhere, an innovative workshop demands an innovative space. Where possible, use a space that is both comfortable for the participants but also lends itself to active, out-loud thinking.

The ideal campsite has multiple spaces. Over the course of a design camp, participants can move between these spaces to create an optimised and inspiring set of experiences for people as they contribute to the emerging outputs.

Sometimes it's more beneficial to hold the design camp in a space that inspires through some connection to the conversation. For example, if we are working on a project related to an architectural piece or a museum exhibition, it makes more sense to hold the design camp *in situ* so that participants can be naturally inspired through observation and immersion.

Solution blueprinting

Time needed	1 week to 1 month. Note that a rapid solution blueprint can be developed in a few days if an initial cut is needed quickly.
Design System phase	Innovate / Formulate / Implement / Recommend

There are many different types of blueprint. Some provide a detailed design of a service interaction or an organisation. Some are more strategic about how different stakeholders will work together, for example in a health, energy or regulatory system. A blueprint comprises four layers:

- the strategy layer, which sets out the purpose, direction and objectives for the design and might also include the constraints and the multi-stakeholder needs

- the human experience layer, showing how a group of people experience the system

- a layer describing the products, service and channels that will shape a person's experience

- the delivery layer, describing the roles and skills of people, the flows of work, the IT systems and any other elements that need to be in place to deliver the experience.

Blueprints define all the broad aspects of a change without defining all of the finer details. Instead, they can focus on a range of concepts, from new organisational structures to new products, channels, IT systems or even strategic visions of a whole organisation. A blueprint should envision how the design will affect the social, economic or environmental system and guide the detailed design, living on beyond the scope of the project.

How-to

While each blueprint will be different, the guidelines below will help you structure your thinking. The best approach is a co-design process engaging diverse stakeholders.

- Start with a clear understanding of the intent of a change.

- Use a core design team to bring in a range of experience, knowledge and viewpoints and collaboratively design the concepts.

- Have a sense of experimentation, an openness to failing and an enthusiasm for iteration.

- Start fuzzy, using divergent thinking but end on something clear and convergent.

- Ensure that the design forces diligent consideration of the impact of the design on all stakeholders, including those in the wider environment.

- Actively seek out potential unintended consequences.

Keep in mind

- For service design projects, the blueprint is a key design output. For an organisational design project, a blueprint is also often a key output. When moving from designing the solution to implementing the solution, use a blueprint to seek endorsement and collective support for a way forward.

- A good service blueprint will have a strong explanation of the 'who' (the client, the personas and so on). Blueprints reflect our core belief that people are at the centre of effective service delivery and organisational change.

- There can be a tendency to design a blueprint in the same order as the design process. However, the reader often wants the information in the opposite order. Therefore design a blueprint in reverse. Start by asking what the change is people want to see. Then, move to discuss how the change will work. Conclude with the justification for why this is the preferred design.

Figure 33: Design blueprint for ICT project in rural Ghana

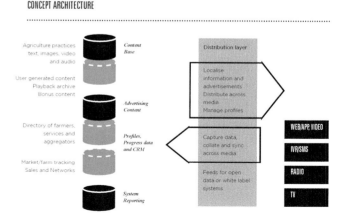

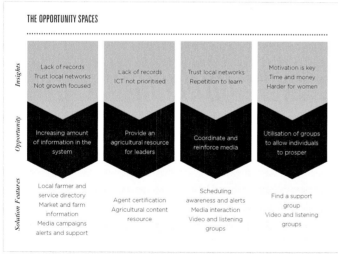

BUILDING AND TESTING

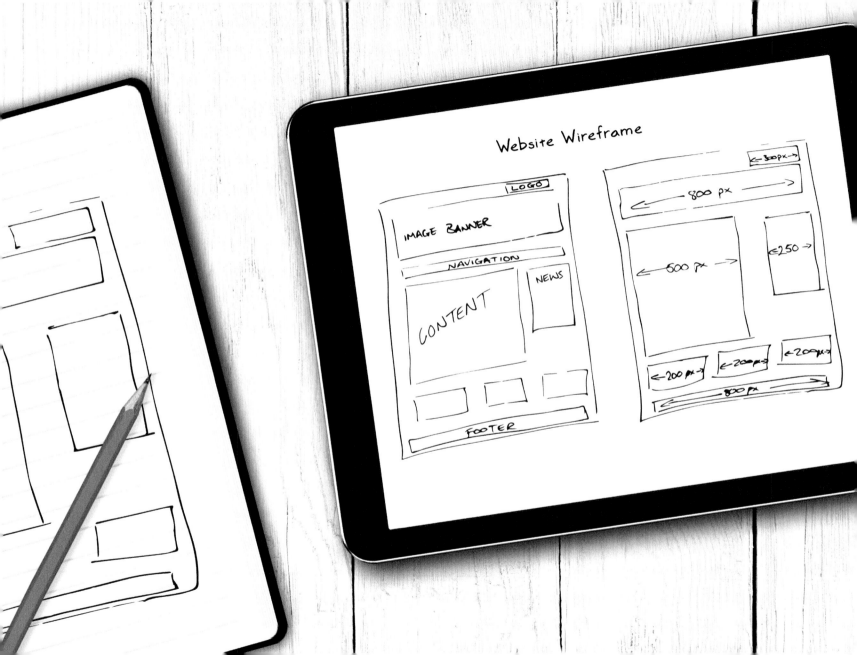

Prototyping

Time needed	Multi-day fieldwork, user testing or workshops
Design System phase	Innovate
You'll also need	Sticky notes
	Tack
	Flipchart
	Marker pens
	Camera / video camera
	Modelling clay
	Tape
	Foam core / cardboard

Prototyping is a design technique that arrives at a solution through the progressive refinement of a working model of a concept. The first prototype will be very basic; later versions will be quite sophisticated.

The basic concept is to 'make to learn', reinforced by a mindset of experimentation, a bias toward action and encouragement to show rather than tell. The purpose is to rapidly test various aspects of a design, socialise an idea in a physical way, illustrate ideas or features, and gather early feedback.

Low-fidelity prototypes (lowest level of sophistication, though still able to be tested) help to validate concepts, spark ideas and inspire further thinking. They make ideas visible, touchable and malleable. Medium- to high-fidelity prototypes (reaching a higher level of sophistication that can be engaged in more in-depth user testing) test and communicate a suggested design. Importantly, prototypes should allow people to get physical fast, prioritising speed and ease while keeping the level of risk low. Producing something rough and basic allows participants to feel more comfortable scrutinising and making alterations.

Prototyping embodies the essence of design thinking and practice – it's all about moving quickly to learn what a new future state can be. Prototyping can be considered a risk-management technique because the experimentation process is frontloaded, when costs of change are low. It's much costlier to change projects that are well advanced. Prototyping increases the quality of solutions and the likelihood of achieving a better solution.

How-to

Figure 34 describes the stages of prototyping, from exploring an idea to specifying a representation of a solution.

Prototyping can take many forms. To determine the best technique for your situation, ask:

- **What are we trying to achieve?** What is the output that we need to make? For example, mock-ups showing how a touchpoint can be used or conceptual diagrams that explain the scope of the product or service?

- **What do we know?** What research insights exist? What is the context of the problem/opportunity? What are the user requirements? For example, consider the variation in customer needs and expectations.

- **What do we want to learn?** What works well and what won't work in our context? Are there any opportunities that we are missing? How can connections be shown? What flow of information is useful and what form might it take? For example, decide whether to work online or on paper.

When prototyping, consider using one or more of the following techniques:

- Storyboarding – Play out a prototype frame by frame. This technique is especially useful if you are designing an interaction or a new strategy rather than a physical product. What does a scene from the ideal future state look like?

- Scripting – Sometimes a good place to start when prototyping is to have the story of a user's experience clear in your head first. What is the narrative of the design product? What is the scenario in which you visualise the ideal user experience?

- Role play – This means putting yourself in the shoes of the user. Get into character and better understand a user's experience with the prototype.

- Re-enactment – Dramatisations of events. This can be related to customer journey maps, ethnographic research or hypotheticals.

- Frankenprototypes – 'Quick and dirty' mock-ups of a design product.

- Sketch-over scenarios – These start with photos or sketches of a scene. Participants sketch over the scene and create alternative scenarios. This technique is particularly helpful for designers to understand placement, positioning and levelling.

- Modelling – Creating scaled representations of scenes, actors, products and so on. Modelling can be very useful in terms of playing out different scenarios.

- In-context – The strength of prototyping in-context is that designers can test the more candid aspects of the prototype. Let's say you're building something that is for public use, such as a menu or a visitor's information map. It can be very illustrative to test the prototype in its own environment with users, because the designer can see what kind of attention the prototype gets, who has the opportunity to interact with it, and so on. When seeing how a prototype works *in situ*, we tend to use natural speech and have real-time actions and reactions.

Keep in mind

Prototyping is a 'show not tell' technique, so have enough supplies available for people to create at least a basic version of their solution. Prototypes are there for us to learn from, not to prove something. It's important to avoid a situation where a participant's thinking is interrupted by not having the right tools with which to translate their thoughts into something real and physical.

> "When prototyping, sometimes you want to validate your own opinions so much that you're unable to keep listening and learning from others' input. Strive to avoid that. Don't get emotionally bound to your ideas, but keep it light and allow others to build on your concepts and opinions."
>
> *Jim Scully*
> Partner and Managing Director, ThinkPlace New Zealand

Figure 34: Prototyping stages

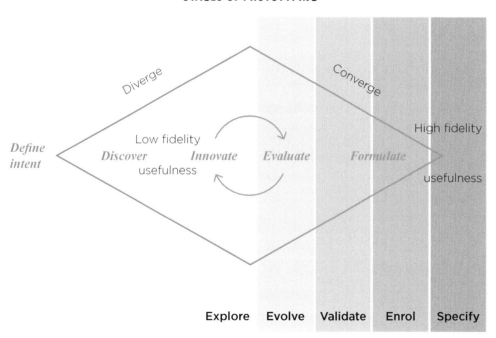

STAGES OF PROTOTYPING

Explore: The primary purpose is to make ideas tangible and to get feedback quickly.

Evolve: Once the fundamental value proposition is understood and the general arc of the concept is identified, identify variations that can extend the value of the concept.

Validate: Once concepts have been evolved and broadly evaluated, create something that has higher fidelity but respects the principle of economy. If validation fails, begin again – there should still be enough flexibility to learn and re-direct the process.

Enrol: This is about socialising the vision of the future – especially useful when the project involves significant change.

Specify: Provide a representation of a solution (more than text-based) and a description of the solution for the design.

Usability testing

Time needed	Varies based on product tested
Design System phase	Innovate
You'll also need	Usability protocol
	Usability schedule
	Usability scenario
	Researched script
	Participant consent form

Usability testing is a technique that evaluates the viability, desirability and possibility of a product or service by testing it on users. It is one of the many ways to ensure that prototyping and testing techniques keep the user experience at the heart of the process. There are three types of usability testing:

- explorative – early testing to understand the user's mental model and expectations

- evaluative – tests user experience, effectiveness and overall satisfaction when using the product

- comparative – tests two or more products to determine user preference.

While in-depth user research helps us to look at individual user experiences, usability testing allows us to capture patterns across a large group of people, of which the individual users themselves may not be aware. We typically do usability testing when we are trying to 'converge' on a solution, though general usability applies to every step in a project (for example, through 'paper and pen prototyping'). Results from usability testing come in the form of both quantitative measures (such as task success, metrics) and qualitative measures (such as the discovery and prioritisation of issues).

By performing usability testing, we bring the user's voice into the process early and often. This ends up being a more cost-effective way of testing for the client and yields a stronger, more user-centred product. We understand that a product can't be developed without any input from the users. Usability testing not only provides the feedback right away – you're never quite sure what you'll find! – it provides feedback that is specific to the needs of the users. Usability testing also allows you to introduce the role of an observer, whether it's the client or an outside observer, which helps to remove the idea that we are the expert on the product. Instead, the focus remains on the user experience.

How-to

To conduct a usability test:

1. Design – Clarify the outcome of the research, assign roles, identify the user cohorts and develop the test script.

2. Schedule – Confirm test details (day, place, time) and recruit users.

3. Prepare – Finalise research materials and set-up of the usability lab.

4. Test – Conduct the usability testing, collecting qualitative and quantitative data.

5. Synthesise – Make sense of the data, drawing themes and patterns of user experience and developing a solution.

Usability itself is measured in several ways. All of these measurement types should be used for a robust evaluation:

- Self-report – These include statements, comments and answers to probes by the researcher. This verbatim data is essential to the synthesis stage – there's nothing quite as powerful as the words of a user.

- Observed behaviour – This requires skill from the researcher. There's much to be learnt from unconscious and subtle behaviour that can be useful. Sometimes the behaviour can even contradict what is said.

- Rating – Ask participants to rate their satisfaction or confidence. This provides a numerical value that can be compared within and between participants, and also provides another opportunity to probe. This can be used throughout the task to measure end-to-end experience, as well as for elements of the product, such as individual pages of a website.

- Facts and figures – Quantitative data rounds out the measurement data. Metrics could include the time it takes to complete a task, the number of mouse clicks, or eye/mouse hover patterns. Collecting this data requires a more sophisticated set-up.

Keep in mind

When doing usability testing, listen to what people are saying, but pay more attention to what they're doing. Sometimes these two things don't align because the user may not have the technical knowledge to articulate their actions or pain points. It's the designer's job to distinguish the real, underlying message the user is trying to communicate.

Researchers should try to come together to download their findings and insights as soon as possible in the field, for example during breaks or at the end of the day.

For some user cohorts, such as people who are difficult to reach or vulnerable, a professional researcher may not be best placed to conduct testing. It may be preferable to engage frontline staff or community services operating in the local area to be trained and supported to help with usability testing.

Don't underestimate the time and effort needed for recruitment.

"Usability testing lets you say: neither we nor the client are experts. Let's watch what people do and look at it objectively."

Charlie Mere
Senior Executive Designer, ThinkPlace Australia

THINKPLACE
about the future
of Canberra.

CHANGE THE WAY YOU THINK

OUTLOOK

OUTLOOK

In this book we've sought not only to explain how to design in the context of complex systems, but also to provoke thinking about the ways design can cut through complexity to create real public, shared or collective value.

Starting with an explanation of the principles of co-design and innovation that guide design in complex systems, we looked at the operative concepts and theoretical models that bind the methods together. These underpin the core methodology that guides the designer through the design process.

The core four-diamond model – our Design System™ – is intentionally straightforward and flexible. Designers should never follow a 'paint by numbers' approach. Every design challenge will be different and require thought about the best detailed approach to follow. The four-diamond model provides a methodological framework that the designer can use to move from the conceptual to the pragmatic. The designer must be a maker, evaluator and implementer. As shown in the four-diamond model, thinking about evaluation starts in the first diamond and is considered throughout the designing and making phases. Designers are experienced evaluators who boldly scrutinise their own design and that of others to find powerful, new opportunity areas.

We then explored how the methodology works in practice, taking a close look at the core areas of design expertise and the suite of tools and techniques that the designer uses to move the design forward.

We're genuine about sharing the Design System. There are so many important and worthwhile complex design challenges in the world, and we hope that more can be tackled through design.

Now that we've given you a glimpse into the approach that we've developed over the past decade, it's worth considering how our methodology fits into the bigger picture. The Design System has been applied locally to find relevant solutions to problems. It is also being increasingly applied to global level challenges such as health, nutrition and clean energy.

Where do we see the field going into the future? We live in a world of increasingly complex systems and challenges. For fourth-order designers, this means that the 'menu' of wicked problems is ever growing. What were once localised issues are now major, systemic problems that affect us all. We are constantly faced with images and stories of major humanitarian crises, natural disasters, sustainable energy issues, climate change, failing or emerging economies, overthrown

governments, and more. While there is less war and violence in modern times than there has ever been, we have new types of problems to deal with that have different scales of impact. Cybercrime, for instance, may affect a vast number of people in a short amount of time. The sheer magnitude of these problems creates a ripple effect that is at times positive and at other times negative. As people try to understand these problems and solve them, we see an increased range of empathy and interconnectedness.

The United Nations Sustainable Development Goals describe 17 challenges that are universally agreed by the almost 200 signatory countries as issues the world needs to make progress in. Where there are large-scale problems, the number of people who have a common interest in solving that problem is also large. As designers, we've not dealt with shared interest and connectedness on such a scale before.

Technology has a critical role in all of this. Technology accelerates everything we've mentioned and gives people opportunity and an equal voice where previously there was none. Moreover, technological advantage can offset material disadvantage. Technology has the potential to be a great equaliser, allowing all people to make informed decisions and to participate in global commerce.

Online forums can provide safe havens for people who are persecuted or marginalised in some way. Digital commerce platforms and currencies are opening up opportunities for people to participate in areas that were previously well out of reach.

Complementing the rise of technology and the increasing interconnectedness of 'things', there is a growing appetite for collective solutions. People are demanding more opportunities for collaboration with their governments. Communities want the benefits that government regulation brings without the cost and bureaucracy. To meet those needs, public sector organisations are looking for innovative ways to reach those shared outcomes without being intrusive, invasive or burdensome. Similarly, the private sector is becoming more consumer-driven. The balance of power is changing, giving consumers and communities a seat at the table.

So what does all this mean for design? Well, technology doesn't respect global boundaries. With the rapid sharing of information and the enhanced mobility of people in today's world, design moves faster when you can build on others' ideas.

The changing multimedia landscape means we can collaborate more effectively and more easily than ever before. We see governments and other organisations outsourcing creativity, inviting others to design policy, strategies and measurement frameworks. We also see organisations making collaborative modes of working a priority. This means there is a growing capacity for change in our communities. We understand that co-design, co-production and co-creation give people a seat at the table. More than that, however, co-design cuts through complexity and invites people to build collective ownership of a solution.

This is the kind of vision we espouse. This is how we see design in light of a rapidly changing global context: those for whom you are designing should be those with whom you are designing. As designers, we are optimistic and excited about the future.

BIBLIOGRAPHY AND FURTHER READING

Alvesson, M. (2016). *The stupidity paradox*. London: Profile Books.

Beckman, S. and Barry, M. (2007). Innovation as a learning process: Embedding design thinking. *California Management Review*, 50(1).

Bizios, G. (1991). *Architecture reading lists and course outlines*. Durham, NC: Eno River Press.

Body, J. and Forrester, S. (2014). Synthesising policy and practice: The case of co-designing better outcomes for vulnerable families. In: C. Bason, ed., *Design for policy*. London: Ashgate Publishing Group.

Body, J. and Habbal, F. (2015). The innovation ecosystem. In: B. Banerjee and S. Ceri, ed., *Creating innovation leaders: A global perspective*. Springer International Publishing.

Buchanan, R. (1992). Wicked problems in design thinking. *Design Issues*, 8(2), pp. 5–21.

Buchanan, R. (2008). Management and design: Interaction pathways in organizational life. In: R. Boland and F. Collopy, ed., *Managing as design*. Stanford: Stanford Business Books, pp. 54–63.

Bucolo, S. (2015). *Are we there yet? Insights on how to lead design*. Amsterdam: BIS Publishers.

Butcher, J. and Gilchrist, D. (2016). *The three sector solution*. Canberra: ANU Press.

Castelhano, M. and Henderson, J. (2008). The influence of colour on the perception of scene gist. *Journal of Experimental Psychology*, 34(3), pp. 660–75.

Corkindale, G. (2011). The importance of organizational design and structure. *Harvard Business Review*, 89(1–2).

Duarte, N. (2011). *Slide:ology*. Köln: O'Reilly.

Eames, C. (1972). *Design Q&A*. [Short film].

Eames, C. (1998). attr. In: G. Bizios, ed., *Architecture reading lists and course outlines*. Durham, NC: Eno River Press, p. 494.

Eames, C. and Eames, R. (1961). *ECS*. [Film].

Eisenhower, D. (1957). *Address to National Defense Executive Reserve Conference*, Washington D.C.

Gleick, J. (2008). *Chaos*. New York: Penguin Books.

Graeber, D. (2016). *The utopia of rules*. New York: Melville House Publishing.

Gray, D., Brown, S. and Macanufo, J. (2010). *Gamestorming*. Beijing: O'Reilly.

Gregory, R. (1970). *The intelligent eye*. London: Weidenfeld & Nicolson.

Guilford, J. (1967). *The nature of human intelligence*. New York: McGraw-Hill.

Gunn, W., Otto, T. and Smith, R. (2016). *Design anthropology*. London: Bloomsbury Academic.

Hamel, G. and Prahalad, C. (1989). Strategic intent. *Harvard Business Review*, 67(3), pp. 63–76.

Hill, D. (2012). *Dark matter and trojan horses: A strategic design vocabulary*. UK: Strelka Press.

Kahneman, D. (2013). *Thinking, fast and slow*. New York: Farrar, Straus and Giroux.

Kelley, T. and Littman, J. (2005). *The ten faces of innovation*. New York: Currency/Doubleday.

Kolb, D. A. (1984). *Experiential learning: Experience as the source of learning and development* (Vol. 1). Englewood Cliffs, NJ: Prentice-Hall.

Kotter, J. (2012). *Leading change*. Boston: Harvard Business Review Press.

Kotter, J. and Cohen, D. (2002). *The heart of change*. Boston: Harvard Business School Press.

Kurtz, C. and Snowden, D. (2003). The new dynamics of strategy: Sense-making in a complex and complicated world. *IBM Systems Journal*, 42(3), pp. 462–83.

Lidwell, W., Holden, K. and Butler, J. (2010). *Universal principles of design*. Beverly, MA: Rockport Publishers.

Liedtka, J., Ogilvie, T. and Brozenske, R. (2014). *The designing for growth field book*. New York: Columbia Business School Publishing.

Martin, B. and Hanington, B. (2012). *Universal methods of design*. Beverly, MA: Rockport Publishers.

McCloud, S. and Martin, M. (2014). *Understanding comics*. New York: William Morrow/HarperCollinsPublishers.

Moore, M. (1995). *Creating public value: Strategic management in government*. Cambridge, MA: Harvard University Press.

Moore, M. (2002). *Creating public value*. Cambridge, MA: Harvard University Press.

Neumeier, M. (2009). *The designful company*. Berkeley: New Riders.

Owen, H. (1999). *The spirit of leadership: Liberating the leader in each of us*. San Francisco: Berrett-Koehler.

Owen, H. (2008). *Open space technology*. San Francisco: Berrett-Koehler Publishers.

Phills, J., Deiglmeier, K. and Miller, D. (2008). Rediscovering social innovation. *Stanford Social Innovation Review*, 6(4).

Popper, K. (1991). Of clouds and clocks: An approach to the problem of rationality and the freedom of man. In: P. Meehl, D. Cicchetti and W. Grove, ed., *Thinking clearly about psychology V1: Matters of public interest*. Minneapolis: University of Minnesota Press, pp. 100–139.

Potter, N. (2009). *What is a designer*. London: Hyphen Press.

Prahalad, C. and Ramaswamy, V. (2000). Co-opting customer competence. *Harvard Business Review*, 78(1).

Ralph, J., Allert, R. and Joss, B. (1999). *Review of business taxation: A tax system redesigned*. Canberra: Australian Department of the Treasury.

Rittel, H. and Webber, M. (1973). Dilemmas in a general theory of planning. *Policy Sciences*, 4(2), pp. 155–69.

Sanders, E. and Stappers, P. (2016). *Convivial toolbox*. Amsterdam: BIS Publishers.

Sanoki, T. and Sulman, N. (2011). Colour relations increase the capacity of visual short-term memory. *Perception*, 40(6), pp. 635–48.

Seelig, T. (2012). *InGenius*. London: Hay House.

Senge, P. (2006). *The fifth discipline*. London: Random House Business Books.

Shostack, G. (1984). Designing services that deliver. *Harvard Business Review*, 62(1), pp. 133–9.

Simon, H. (1988). The science of design: Creating the artificial. *Design Issues*, 4(1/2), pp. 67–82.

Snowden, D. (2002). Complex acts of knowing: Paradox and descriptive self awareness. *Journal of Knowledge Management*, 6(2), pp. 100–11.

Snowden, D. and Boone, M. (2007). A leader's framework for decision making. *Harvard Business Review*, 85(11), pp. 68–76.

Squires, S., Byrne, B. and Sherry, J. (2002). *Creating breakthrough ideas*. Westport, CT: Bergin & Garvey.

Stephenson, W. (1953). *The study of behavior*. Chicago: University of Chicago Press.

Terrey, N. and Evans, M. (2016). Design thinking and public policy reform. In: S. Gerry and M. Evans, ed., *Evidence-based policy making in the social sciences*. Bristol, UK: Policy Press.

Tomasi di Lampedusa, G. (1958). *The leopard*. New York: Pantheon Books.

Tufte, E. (2010). *Beautiful evidence*. Cheshire, CT: Graphics Press.

Waldrop, M. (2008). *Complexity*. New York: Simon & Schuster.

Young, I. (2011). *Mental models*. Sebastopol: Rosenfeld Media.

INDEX